Calkin Algebras and Algebras of Operators on Banach Spaces

Lecture Notes in Pure and Applied Mathematics

1. *N. Jacobson,* Exceptional Lie Algebras
2. *L.-Å. Lindahl* and *F. Poulsen,* Thin Sets in Harmonic Analysis
3. *I. Satake,* Classification Theory of Semi-Simple Algebraic Groups
4. *F. Hirzebruch, W. D. Neumann,* and *S. S. Koh,* Differentiable Manifolds and Quadratic Forms
5. *I. Chavel,* Riemannian Symmetric Spaces of Rank One
6. *R. B. Burckel,* Characterization of C(X) among Its Subalgebras
7. *B. R. McDonald, A. R. Magid,* and *K. C. Smith,* Ring Theory: Proceedings of the Oklahoma Conference
8. *Yum-Tong Siu,* Techniques of Extension of Analytic Objects
9. *S. R. Caradus, W. E. Pfaffenberger, and Bertram Yood,* Calkin Algebras and Algebras of Operators on Banach Spaces

Other volumes in preparation

Calkin Algebras and Algebras of Operators on Banach Spaces

S. R. Caradus
QUEEN'S UNIVERSITY

W. E. Pfaffenberger
UNIVERSITY OF VICTORIA

Bertram Yood
PENNSYLVANIA STATE UNIVERSITY

MARCEL DEKKER, INC. New York 1974

MARCEL DEKKER, INC.

270 Madison Avenue, New York, New York 10016

LIBRARY OF CONGRESS CATALOG CARD NUMBER: 74-15630

ISBN: 0-8247-6246-0

Current printing (last digit):
10 9 8 7 6 5 4 3 2 1

PRINTED IN THE UNITED STATES OF AMERICA

This book is dedicated to

I. C. Gohberg

for his important and lasting
contributions to functional analysis.

PREFACE

The interaction between the theories of Banach algebras with involution and that of bounded linear operators on a Hilbert space has been extensively developed ever since the birth of Banach algebra theory. In the meantime there have evolved, in a natural way but at a much slower pace, interesting connections of Banach algebras with the theory of bounded linear operators on a Banach space. These notes are intended to provide an introduction to the latter set of ideas. Here central items of interest include Fredholm operators, semi-Fredholm operators, Riesz operators and Calkin algebras.

We begin with a treatment of the classical Riesz-Schauder theory which takes advantage of more recent developments. Some of this material (Riesz-Schauder operators) appears here for the first time. In order to make our exposition suitable for readers with a modest background, we have included an introductory chapter on Banach algebras. With admirable restraint we have not tried to give a short course in Banach algebras but have included only material rather directly relevant to our aims. This is followed by chapters on Riesz and semi-Fredholm operators.

Let $B(X)$ be the Banach algebra of all bounded linear operators on a Banach space X. Chapter 5 features the remarkable result of Gohberg, Markus and Fel'dman (1960) that, for $X = \ell_p$, $1 \leq p < \infty$ and $X = c_o$, the compact operators on X furnish the only proper closed two-sided ideal in $B(X)$. (The case $X = \ell_2$ is due to Calkin (1941)). Finally in Chapter 6 we indicate relations of our subject matter to a variety of recent developments.

The authors are indebted to the National Science Foundation, the National Research Council of Canada and the University of Victoria (faculty research grant) for financial support. We are also very grateful to the typist, Mrs. D. J. Leeming, and to the publishers for their assistance and patience.

CONTENTS

Calkin Algebras and Algebras of Operators on Banach Spaces

Chapter 1
INTRODUCTION

We assume that the reader is somewhat familiar with the rudiments of functional analysis as developed in the classical treatise of Banach [2] or in later expositions such as Part I of the encyclopaedic work of Dunford and Schwartz [22]. Let X and Y be two Banach spaces (over either the real field or complex field, but the latter will be much more interesting for us later). We denote by $B(X,Y)$ the Banach space of all bounded linear transformations T of X into Y with the norm

$$||T|| = \sup_{||x||=1} ||T(x)||.$$

In case $X = Y$ we write $B(X)$ in place of $B(X,X)$. $B(X)$ is a Banach algebra (see Chapter 2) under the usual definitions for operator addition and multiplication.

A linear transformation $T \varepsilon B(X,Y)$ is called <u>compact</u> (or <u>completely continuous</u>) if $T(S)$ has compact closure in Y where S is the unit ball of X. Often it is convenient to use the equivalent definition which asserts that T is compact if and only if, for each bounded sequence $\{x_n\}$ in X, there exists a subsequence $\{x_{n_k}\}$ and an element $y \varepsilon Y$ such that $T(x_{n_k}) \rightarrow y$. Another equivalent formulation is that the image, under T, of a bounded set in X is totally bounded in Y.

Let $K(X,Y)$ denote the set of all compact linear transformations in $B(X,Y)$ and write $K(X,X)$ as $K(X)$. Elementary arguments, which we omit, show that $K(X,Y)$ is a closed linear subspace in $B(X,Y)$ and $K(X)$ is a closed two-sided ideal in $B(X)$ (see [22, p. 486]). This fact enables us to define the <u>Calkin algebra over X</u> as the quotient algebra $C(X) = B(X)/K(X)$. $C(X)$ is itself a Banach algebra (see Chapter 2) in the quotient algebra norm

$$||T + K(X)|| = \inf_{U \varepsilon K(X)} ||T + U||.$$

We shall use π to denote the natural homomorphism of $B(X)$ onto $C(X)$; $\pi(T) = T + K(X)$.

Calkin algebras were first studied seriously by Calkin [7] in the special but very interesting case where X is a separable Hilbert space. The treatment of Calkin algebras over Banach spaces was begun by Yood in [69]. Further references will be given as the theory is developed. This first introductory chapter is devoted to a modernized version of what is now referred to as the "Riesz-Schauder theory." The original paper by F. Riesz [55], although now fifty-six years old, can still be warmly recommended to all readers. This pioneering paper is unusual for its elegance and charm. It also contains a careful account of the motivation of the theory in its relevance to the theory of integral equations.

To get at this connection in its simplest form we consider $C[a,b]$, the Banach space of all complex-valued continuous functions on the bounded closed interval $[a,b]$, under the sup norm,

$$||f|| = \sup_{t\varepsilon[a,b]} |f(t)|.$$

A continuous complex-valued function $K(s,t)$ defined on $[a,b] \times [a,b]$ gives rise to a bounded linear operator T on $C[a,b]$ via the rule

$$F(f)(s) = \int_a^b K(s,t)f(t)dt.$$

Given $\varepsilon > 0$ there corresponds, by the uniform continuity of $K(s,t)$, a number $\delta > 0$ such that

$$|K(s_1,t) - K(s_2,t)| < \varepsilon, \quad s_1, s_2, t \ \varepsilon \ [a,b]$$

provided that $|s_1 - s_2| < \delta$. For such s_1 and s_2, we see that

$$|T(f)(s_1) - T(f)(s_2)| < \varepsilon \ (b - a)||f||.$$

This shows that the functions $T(f)$, for f in the unit ball S, of $C[a,b]$, are equi-continuous as well as bounded in norm. A classical theorem of Arzelà asserts that $T(S)$ is totally bounded in $C[a,b]$. Consequently T is a compact linear operator.

The classical Fredholm integral equation is

$$\lambda f(s) - \int_a^b K(s,t)f(t)dt = g(s), \quad a \leq s \leq b$$

where g is given in $C[a,b]$, λ is a parameter and f is the unknown. Using I to be the identity operator on $C[a,b]$ we can recast this equation into the form $(\lambda I - T)f = g$. Thus, following Riesz, we are naturally led to the study of operators of the form $V = \lambda I - T$ on any Banach space X where $\lambda \neq 0$ and $T \in K(X)$.

The Riesz-Schauder theory concentrates attention on these operators of the form $V = \lambda I - T$, $\lambda \neq 0$ and T compact. Their results have been considerably generalized and the original arguments are obsolete. This comes about in part because the operators V are special cases of a class of operators called Fredholm operators (and these are in turn special cases of semi-Fredholm operators). Moreover compact operators T are special cases of a class of operators called Riesz operators. From a study of these classes of operators one obtains all the properties of the operator $\lambda I - T$ and much more. In this chapter we approach the Riesz-Schauder theory via the theory of semi-Fredholm operators. We start off in a modest way with two elementary lemmas of Riesz [55].

(1.1.1) LEMMA

Let E be a proper closed linear subspace in a normed linear space Z. Let $0 < \varepsilon < 1$. Then there exists $x_0 \in Z$, $||x_0|| = 1$ where $\text{dist}(x_0,E) \geqq 1 - \varepsilon$.

Proof. First select $y \in Z$ where $y \notin E$. Let $d > 0$ denote the distance of y from E. There exists $u \in E$ such that

$$||y - u|| < \frac{d}{1 - \varepsilon} .$$

Then $x_0 = c(y - u)$, where $c = \dfrac{1}{||y - u||}$, has the desired properties. For if we take any $w \in E$, we get

$$||x_0 - w|| = c||y-(u+||y-u||w)|| \geqq c\ d > 1 - \varepsilon. \blacksquare$$

(1.1.2) LEMMA

Let B be the unit ball in a normal linear space Z. The following statements are equivalent:

(1) Z is finite dimensional

(2) B is compact

(3) B is totally bounded.

Proof. It is sufficient to show that (3) implies (1). Suppose (3) holds, but that (1) fails. Select x_1 with $||x_1|| = 1$. We construct a sequence $\{x_n\}$, by induction, as follows. Suppose that $x_1, \ldots, x_n$ have been selected. Let L_n be the subspace of Z spanned by $x_1, \ldots, x_n$. The preceeding lemma shows that there exists $x_{n+1} \in Z$ where $||x_{n+1} - x_j|| \geqq \frac{1}{2}$ for $j = 1, \ldots, n$ and $||x_{n+1}|| = 1$. Therefore $||x_i - x_j|| \geqq \frac{1}{2}$ for $i \neq j$, contrary to (3).■

1.2 On duality

Let $X^*(Y^*)$ denote the conjugate space of the Banach space $X(Y)$. We recall the definition of the adjoint T^* of the linear operator T in $B(X,Y)$. For each $y^* \in Y^*$ and $x \in X$, set

$$T^*(y^*)(x) = y^*[T(x)].$$

One readily checks (as in [22, p. 478]) that $T^* \in B(Y^*,X^*)$ and, in fact, $||T|| = ||T^*||$. It is vital to consider also the adjoint T^{**} of T^*; $T^{**} \in B(X^{**},Y^{**})$. Let $J_X(J_Y)$ denote the canonical embedding of $X(Y)$ into $X^{**}(Y^{**})$ where, for example, given $x \in X$

$$J_X(x)(x^*) = x^*(x)$$

for all $x^* \in X^*$. One verifies directly that J_X is norm-preserving and

$$T^{**}J_X = J_YT.$$

Therefore T^{**} may be viewed as an extension of T if X and Y are canonically embedded in their second conjugate spaces. The elementary details referred to here can be found in any standard text on functional analysis. Duality theory relates properties of X to those of X^* and properties of T to those of T^*. This topic is also crucial and fascinating in the more general setting of topological vector spaces ([57]) but we confine our attention to a Banach space setting.

The following important and elegant theorem of Schauder concerning the relation of T to T^* added much to the exposition of the Riesz theory and to further developments. See [60] and [22, Theorem VI, 5.2].

(1.2.1) THEOREM

Let $T \in B(X,Y)$. Then T is compact if and only if T^* is compact.

Proof. Suppose that T is compact and let $S(S')$ be the unit ball in $X(Y^*)$. We must show that, given $\varepsilon > 0$ there exists a finite subset $y_1^*,\ldots,y_p^*$ of S' such that, to each $y^* \in S'$ there corresponds some y_r^*, $r = 1,\ldots,p$ where $\|T^*(y^*) - T^*(y_r^*)\| < \varepsilon$.

To this end we observe that, as T is compact there is a finite subset $x_1,\ldots,x_n$ of S where, to each $x \in S$, there corresponds some x_j, $j = 1,\ldots,n$ with $\|T(x) - T(x_j)\| < \varepsilon/3$. Consider the mapping W of Y^* into n-dimensional space $\mathbb{C}^n$ defined by

$$W(y^*) = (y^*[T(x_1)],\ldots,y^*[T(x_n)]).$$

Clearly $W \in K(Y^*,C^n)$. Consequently there is a finite set $y_1^*,\ldots,y_p^*$ in S' such that, to each $y^* \in S'$ there corresponds some y_r^*, $r = 1,\ldots,p$ where $\|W(y^*) - W(y_r^*)\| < \varepsilon/3$. We shall see that $y_1^*,\ldots,y_p^*$ is the subset of S' which we desire. That is, for the y^* and y_r^* just discussed, we show that $\|T^*(y^*) - T^*(y_r^*)\| < \varepsilon$.

First observe that, by the definition of W, we have

$$|y^*[T(x_j)] - y_r^*[T(x_j)]| < \varepsilon/3, \quad j = 1,\ldots,n.$$

Let $x \in S$ and select x_j, $j = 1,\ldots,n$ where $\|T(x) - T(x_j)\| < \varepsilon/3$. Then

$$|T^*(y^*)(x) - T^*(y_r^*)(x)|$$
$$\leqq |y^*[T(x) - T(x_j)]| + |y^*[T(x_j)] - y_r^*[T(x_j)]| + |y_r^*[T(x_j) - T(x)]|.$$

Inasmuch as $\|y^*\| \leqq 1$ and $\|y_r^*\| \leqq 1$ we see that

$$|T^*(y^*)(x) - T^*(y_r^*)(x)| \leqq \varepsilon/3 + 2\|T(x_j) - T(x)\| < \varepsilon.$$

Since x is an arbitrary element of S we have our desired inequality.

Now suppose that T^* is compact. Then so is T^{**}. Inasmuch as T^{**} is an extension of T, it follows that T is compact.■

(1.2.2) NOTATION

For a set S in the Banach space X and a set W in X^* we set

$$S^\perp = \{x^* \in X^* | x^*(x) = 0 \text{ for all } x \in S\} \text{ and}$$
$$^\perp W = \{x \in X | x^*(x) = 0 \text{ for all } x^* \in W\}.$$

For $T \in B(X,Y)$ we let $N(T)$ denote the kernel of T, $N(T) = \{x \in X | T(x) = 0\}$ and we let $R(T)$ denote the range of T, $R(T) = \{y \in Y | T(x) = y$ for some $x \in X\}$.

The following two theorems from the theory of linear transformations are basic to our work. Proofs of these standard theorems (known to Banach [2]) would take us a bit far afield and so we omit them here. The reader is referred to the careful treatment in [22, Chapter VI].

(1.2.3) THEOREM

Suppose that, for $T \in B(X,Y)$, $R(T)$ is closed. Then $R(T^*)$ is closed and, in fact, $R(T^*) = N(T)^{\perp}$.

(1.2.4) THEOREM

Suppose that, for $T \in B(X,Y)$, $R(T^*)$ is closed. Then $R(T)$ is closed and, in fact $R(T) = {}^{\perp}[N(T^*)]$.

For a closed linear subspace E of a Banach space X let σ denote the natural homomorphism of X onto the quotient space X/E. As $R(\sigma) = X/E$, σ^* is surely one-to-one. Theorem 1.2.3 shows that σ^* is a bicontinuous isomorphism of $(X/E)^*$ onto $E^{\perp}$. Actually a straightforward calculation shows that σ^* is norm-preserving.

(1.2.5) LEMMA

For a closed linear subspace E of X, E^* is isometrically isomorphic to $X^*/E^{\perp}$.

Proof. For $x_1^*, x_2^* \in X^*$, x_1^* and x_2^* agree on E if and only if $x_1^* - x_2^* \in E^{\perp}$. Therefore, if we define $\tau(x^* + E^{\perp})$ to be the restriction of x^* to the domain of definition E, τ is a well-defined linear mapping of $X^*/E^{\perp}$ into E^*. The Hahn-Banach theorem assures that τ maps $X^*/E^{\perp}$ onto all of E^*.

Any $y^* \in x^* + E^{\perp}$ is an extension of $\tau(x^* + E^{\perp})$. Consequently $||\tau(x^* + E^{\perp})|| \leqq ||y^*||$. By the definition of the norm of a coset as the infimum of the norms of the elements in it we see that

$$||\tau(x^* + E^{\perp})|| \leqq ||x^* + E^{\perp}||.$$

On the other hand, by the Hahn-Banach theorem, $\tau(x^* + E^{\perp})$ has an extension $w^* \in X^*$ with the same norm. Since $w^* - w^* \in E^{\perp}$ we get

$$||x^* + E^{\perp}|| \leqq ||w^*|| = ||\tau(x^* + E^{\perp})||.$$

Therefore τ is an isometry.■

(1.2.6) NOTATION

For $T \in B(X,Y)$, the dimension of $N(T)$ shall be denoted by $\alpha(T)$. (We write $\alpha(T) = \infty$ if $\alpha(T)$ is not finite). If $R(T)$ is closed in Y we let $\beta(T)$ be the dimension of $Y/R(T)$. Again we set $\beta(T) = \infty$ if $Y/R(T)$ is not finite-dimensional.

A result of Kato shown below (Corollary 3.2.5) asserts that $R(T)$ is closed in Y if the linear space $Y/R(T)$ is finite-dimensional. For the time being we avoid a discussion of the space $Y/R(T)$ if $R(T)$ is not closed (see section 4 of Chapter 4).

(1.2.7) PROPOSITION

If $R(T)$ is closed then $\alpha(T^*) = \beta(T)$, $\alpha(T) = \beta(T^*)$ and $\alpha(T) = \alpha(T^{**})$.

Proof. Theorem 1.2.3 asserts that $R(T^*) = N(T)^{\perp}$. By Lemma 1.2.5, $[N(T)]^*$ is isomorphic to $X^*/R(T^*)$. Therefore $\alpha(T) = \beta(T^*)$. As noted earlier Theorem 1.2.3 shows that $[Y/R(T)]^*$ is isomorphic to $R(T)^{\perp}$. Inasmuch as $T^*(y^*) = 0$ if and only if y^* vanishes on $R(T)$ we see that $\beta(T) = \alpha(T^*)$. Then we have $\alpha(T) = \beta(T^*) = \alpha(T^{**})$.■

1.3 Semi-Fredholm operators

(1.3.1) DEFINITION

If $T \in B(X)$ and $R(T)$ is closed, we say that T is a semi-Fredholm operator if either $\alpha(T) < \infty$ or $\beta(T) < \infty$.

In fact we distinguish between two classes of semi-Fredholm operators by the notation

$$\Phi_+(X) = \{T \in B(X) \mid R(T) \text{ is closed and } \alpha(T) < \infty\}$$

and

$$\Phi_-(X) = \{T \in B(X) \mid R(T) \text{ is closed and } \beta(T) < \infty\}.$$

We also set $\Phi(X) = \Phi_+(X) \cap \Phi_-(X)$ and call this the set of Fredholm operators on X (see Chapter 3 for a detailed treatment).

From the duality theory of section 1.2 it follows that if T is a semi-Fredholm operator then

$$\alpha(T) = \beta(T^*)$$

and

$$\beta(T) = \alpha(T^*).$$

Moreover

$$T \in \Phi_+(X) \text{ if and only if } T^* \in \Phi_-(X)$$

and

$$T \in \Phi_-(X) \text{ if and only if } T^* \in \Phi_+(X^*).$$

We begin with a characterization of $\Phi_+(X)$ (taken from [68]).

(1.3.2) THEOREM

Let $T \in B(X)$. $T \in \Phi_+(X)$ if and only if given any bounded set $E \subset X$ which is not totally bounded it is true that $T(E)$ is not totally bounded.

Proof. ($\leftarrow$) Assume T satisfies the second condition in the statement of the theorem. To prove $T \in \Phi_+(X)$ we first show that $N(T)$ is finite dimensional.

Let S denote the unit ball of $N(T)$. If $N(T)$ is not finite dimensional, then S is not totally bounded, so by hypothesis $T(S)$ is not totally bounded. This is a contradiction since $T(S) = \{0\}$. Therefore $\alpha(T) < \infty$.

To see that $R(T)$ is closed let $X = N(T) + W$ where W is a closed linear subspace. Now $R(T) = T(W)$ and $T(W)$ is closed if T has a bounded inverse when restricted to W. But this is true since if T did not have a bounded inverse when restricted to W, there would exist a sequence $\{x_n\}$ in W such that $||T(x_n)|| \to 0$ and $||x_n|| = 1$. So $\{x_n\}$ is totally bounded by our hypothesis on T. Therefore there exists a convergent subsequence $x_{n_k} \to y \in X$. But $T(x_{n_k}) \to T(y) = 0$ implies $y \in W \cap N(T)$, so $y = 0$, a contradiction. Therefore $R(T)$ is closed, and $T \in \Phi_+(X)$.

($\rightarrow$) Assume $T \in \Phi_+(X)$. We have $X = N(T) + W$ where W is a closed subspace. T is one to one on W and $T(W) = R(T)$ so T restricted to W is bicontinuous. We must show that, given any bounded sequence $\{x_n\}$ in X with no Cauchy subsequence, then $T(\{x_n\})$ is not totally bounded. So let $\{x_n\}$ be a sequence with no Cauchy subsequence. Write $x_n = u_n + v_n$ with $u_n \in N(T)$ and $v_n \in W$. Also $T(x_n) = T(v_n)$. Suppose $T(x_{n_k})$ is Cauchy, then $T(v_{n_k})$ is Cauchy and, since T_W^{-1} is continuous, we have $T_W^{-1} o T(v_{n_k})$ is Cauchy. Therefore $\{v_{n_k}\}$ is Cauchy, which implies $\{v_{n_k}\}$ is bounded. We were given $\{x_{n_k}\}$ was bounded so we know $\{u_{n_k}\}$ is bounded in the finite dimensional space $N(T)$. Therefore there exists a subsequence $\{u_{n_{k_j}}\}$ which is Cauchy. But since

$x_{n_{k_j}} = u_{n_{k_j}} + v_{n_{k_j}}$, this implies that $\{x_{n_{k_j}}\}$ is Cauchy,

which is a contradiction.■

(1.3.3) COROLLARY

If T_1 and T_2 are in $\Phi_+(X)$, then $T_1T_2 \in \Phi_+(X)$. Also if W_1 and W_2 are in $\Phi_-(X)$, then $W_1W_2 \in \Phi_-(X)$.

Proof. Assume T_1 and T_2 are in $\Phi_-(X)$. Let E be a bounded subset of X which is not totally bounded. $T_2(E)$ is bounded and by Theorem 1.3.2 it is not totally bounded. Similarly $T_1T_2(E) = T_1(T_2(E))$ is not totally bounded. Therefore by Theorem 1.3.2, $T_1T_2 \in \Phi_+(X)$.

Assume W_1 and W_2 are in $\Phi_-(X)$. Then W_1^* and W_2^* are in $\Phi_+(X^*)$, so by the above $W_2^* W_2^* \in \Phi_+(X^*)$. However, this implies that $W_1W_2 \in \Phi_-(X)$.■

The following corollaries of Theorem 1.3.2 are also proved in a straight forward manner.

(1.3.4) COROLLARY

If T_1 and T_2 are in $B(X)$ and $T_2T_1 \in \Phi_+(X)$, then $T_1 \in \Phi_+(X)$.

(1.3.5) COROLLARY

If T_1 and T_2 are in $B(X)$ and $T_2T_1 \in \Phi_-(X)$, then $T_2 \in \Phi_-(X)$.

(1.3.6) COROLLARY

If T_1 and T_2 are in $B(X)$ and $T_2T_1 \in \Phi(X)$, then $T_1 \in \Phi_+(X)$ and $T_2 \in \Phi_-(X)$.

(1.3.7) COROLLARY

(a) $T + U \in \Phi_+(X)$ if $T \in \Phi_+(X)$ and $U \in K(X)$

(b) $T + U \in \Phi_-(X)$ if $T \in \Phi_-(X)$ and $U \in K(X)$

(c) $T + U \in \Phi(X)$ if $T \in \Phi(X)$ and $U \in K(X)$.

Proof. We examine (a). First it is easy to see that if subsets E_1 and E_2 of X are totally bounded so is the set $E_1 \setminus E_2$. Suppose that E is a bounded set in X which is not totally bounded. By Theorem 1.3.2, $T(E)$ is not totally bounded. However $U(E)$ is totally bounded as $U \in K(X)$. Inasmuch as $T(E) \subset (T + U)(E) \setminus U(E)$ our research above demonstrates that $(T + U)(E)$ cannot be totally bounded. Therefore $T + U \in \Phi_+(X)$ by Theorem 1.3.2.

The statement (b) follows from (a) by taking adjoints and using duality theory.■

We apply Corollary 1.3.7 to the Riesz-Schauder case of $V = \lambda I - T$ where $\lambda \neq 0$ and $T \in K(X)$. Clearly $\lambda I \in \Phi(X)$ and therefore $V \in \Phi(X)$.

A celebrated result about V is the "Fredholm alternative theorem" which asserts that V is one-to-one if and only if $R(V) = X$. A more inclusive result from which that theorem follows is that $\alpha(V) = \beta(V)$. This can be shown in more than one way. It is a consequence of the theory of Fredholm operators. It also follows from the theory of ascent and descent to which we now turn.

1.4 On ascent and descent

For our immediate purposes let X be any linear space and V a linear operator on X. Set $V^o = I$, $V^1 = V$, $V^2 = VoV$ etc. Then consider the set inequalities

$$X = R(V^o) \supset R(V^1) \supset R(V^2) \supset \cdots .$$

If there exists an integer n such that $R(V^n) = R(V^{n+1})$ we say that V has finite descent. In that case the smallest such integer n is denoted by $d(V)$. If no such n exists, we set $d(V) = \infty$. Likewise we examine

$$(0) = N(V^o) \subset N(V^1) \subset N(V^2) \subset \cdots .$$

If there exists an integer n such that $N(V^n) = N(V^{n+1})$ we say that V has finite ascent and the smallest such n is denoted by $a(V)$. It is easy to show that if $d(V) = n < \infty$ $(a(V) = n < \infty)$ then $R(V^r)$, $(N(V^r) = N(V^n))$ for all $r \geq n$ (see [64, p. 271]).

(1.4.1) LEMMA

Suppose that $d(V) = 0$ and $a(V) < \infty$. Then $a(V) = 0$.

Proof. Suppose that the conclusion is false. Then there exists $x_1 \varepsilon X$, $x_1 \neq 0$ with $V(x_1) = 0$. Inasmuch as $R(V) = X$ there exists $x_2 \varepsilon X$ with $V(x_2) = x_1$. By induction we define a sequence $\{x_n\}$ in X with $V(x_{n+1}) = x_n$ for each $n \geqq 1$. But then $V^n(x_{n+1}) = x_1$ whereas $V^{n+1}(x_{n+1}) = 0$. Thus $x_{n+1} \varepsilon N(V^{n+1})$ and $x_{n+1} \notin N(V^n)$ for each n, which is contrary to the hypothesis that $a(V) < \infty$.■

(1.4.2) LEMMA

Suppose that $a(V) < \infty$ and $d(V) < \infty$. Then $a(V) = d(V)$.

Proof. Let Z be the linear space which can be written as $R(V^j)$ for all $j \geqq d(V)$. Clearly V when restricted to the domain of definition Z is a linear operator V_o on Z. As a linear operator on Z, V_o enjoys the property that $d(V_o) = 0$. Clearly $a(V_o) < \infty$. Lemma 1.4.1 applies to show that $a(V_o) = 0$ so that V_o is a linear space isomorphism of Z onto Z.

For convenience we write $d = d(V)$. Obviously $N(V^d) \subset N(V^{d+1})$ but we show the reverse set inequality. For consider $x \varepsilon N(V^{d+1})$ and examine $y = V^d(x)$. Now $V(y) = V^{d+1}(x) = 0$. Inasmuch as $y \varepsilon Z$ we see that $y = 0$. Hence $x \varepsilon N(V^d)$. We have thus established that $a(V) \leqq d = d(V)$.

It remains for us to see that $d \leqq a(V)$; clearly we may suppose that $d \geqq 1$. Then there exists $y \varepsilon R(V^{d-1})$ where $y \notin R(V^d)$ and we can write $y = V^{d-1}(x)$. Let $z = V(y) = V^d(x)$. Recall that V and all its iterates are isomorphisms on Z. Therefore there exists $w \varepsilon Z$ with the property that $V^d(w) = z$. Now we examine the element $u = x - w$. First $V^d(u) = V^d(x) - V^d(w) = 0$. Next $V^{d-1}(u) = y - V^{d-1}(w)$. But if we realize that $V^{d-1}(w) \varepsilon V^{d-1}[R(V^d)] = R(V^{2d-1}) \subset R(V^d)$ we see that $V^{d-1}(u) \neq 0$. Therefore $a(V) \geqq d$ as desired.■

(1.4.3) PROPOSITION

Suppose that $d = a(V) = d(V) < \infty$. Then

$$X = R(V^d) \oplus N(V^d).$$

Proof. Let W be the linear operator V restricted to the domain of definition $R(V^d)$. Inasmuch as $R(V^d) = R(V^{d+1})$ we see that W is a surjective mapping of $R(V^d)$ onto $R(V^d)$. Lemma 1.4.1 shows that W is also one-to-one on $R(V^d)$ as is every power of V. Therefore $R(V^d) \cap N(V^d) = \{0\}$.

Consider $f \in X$. There exists $h \in R(V^d)$ such that $V^d(f) = V^{2d}(h)$. Write

$$f = V^d(h) + [f - V^d(h)].$$

We complete the proof by observing that $V^d[f - V^d(h)] = V^d(f) - V^{2d}(h) = 0$.■

Henceforth X shall denote a Banach space.

(1.4.4) DEFINITION

We say that $T \in B(X)$ is a Riesz-Schauder operator if T is a Fredholm operator, $a(T) < \infty$ and $d(T) < \infty$.

(1.4.5) THEOREM

The following statements concerning $V \in B(X)$ are equivalent

(a) V is a Riesz-Schauder operator

(b) V can be written as $V = T + U$ where $T \in B(X)$ is an isomorphism of X onto X, $U \in B(X)$ has finite-dimensional range and $TU = UT$.

(c) V can be written as $V = T + U$ where $T \in B(X)$ is an isomorphism of X onto X, $U \in K(X)$ and $TU = UT$.

Proof. Assume (a). Let m be any odd integer where $m > d$, d being the common value of $a(V)$ and $d(V)$ (see Lemma 1.4.2). By Proposition 1.4.3 we write

$$X = R(V^m) \oplus N(V^m). \qquad (1)$$

Let P be the projection of X onto $N(V^m)$ with null-space $R(V^m)$. $P \in B(X)$ (by the closed graph theorem). As noted in the proof of Lemma 1.4.2, V maps $R(V^m)$ onto $R(V^m)$ and V is one-to-one on $R(V^m)$. Therefore $V^m + P$ is an isomorphism of X onto X.

Inasmuch as $V^m(x) = 0$ if $V^{m+1}(x) = 0$, we see that $x \in N(V^m)$ if and only if $V(x) \in N(V^m)$. Let $u = y + z$ in the decomposition (1). Then

$$PV(u) = PV(y) + PV(z) = V(z) = VP(u), \quad u \in X.$$

Therefore $PV = VP$. Set

$$W = V^{m-1} - V^{m-2} P + V^{m-3} P - \cdots + P$$

where the signs alternate. As m is odd and $PV = VP$ we get

$$V^m + P = (V + P)W = W(V + P).$$

But $V^m + P$ is invertible in $B(X)$. Therefore so is $V + P$ which establishes (b).

Clearly (b) implies (c). Assume (c) where $V = T + U$ as stated. By Corollary 1.3.7 we see that $V \in \Phi(X)$ and, by Corollary 1.3.3, every $V^n \in \Phi(X)$, $n = 1,2,\ldots$. We show that $a(V) < \infty$. For suppose otherwise. By Lemma 1.1.1 there exists, for each $n = 1,2,\ldots$, $x_n \in N(V^{n+1})$, $||x_n|| = 1$ and $||x_n - x|| \geqq \frac{1}{2}$ for all $x \in N(V^n)$.

We shall need the fact (guaranteed by Banach's theorem [22, p. 57]) that, for some $k > 0$, $||T(x)|| \geqq k||x||$ for $x \in X$. Take two positive integers $m > n$. We have

$$U(x_m) - U(x_n) = z - T(x_m) \tag{2}$$

where $z = T(x_n) - V(x_n) + V(x_m)$. Since $TV = VT$ and $V^m(x_n) = 0$, we see that $z \in N(V^m)$. Moreover there exists $y \in X$ such that $T(y) = z$. Furthermore $T^{-1}V = VT^{-1}$. Therefore $V^m(y) = V^mT^{-1}(z) = T^{-1}V^m(z) = 0$. Putting this information together we get, from (2) that

$$||U(x_m) - U(x_n)|| = ||z - T(x_m)|| \geqq k||y - x_m|| \geqq k/2.$$

This is a contradiction as the $\{x_n\}$ constitute a bounded set and U is a compact operator. Therefore $a(V) < \infty$.

Next consider $V^* = T^* + U^*$. By duality theory V^* satisfies (c) as an element of $B(X^*)$. The proof above then gives $a(V^*) < \infty$. Then, by Theorem 1.2.4, we see that $d(V) < \infty$. Consequently (c) implies (a).■

Curiously, the U of (b) can always be chosen to be the negative of an idempotent operator.

The question arises of whether or not every operator of the form $T + U$ (as in (b) or (c)), but where $TU \neq UT$, need be a Riesz-Schauder operator. The answer is negative. For let X be the Banach space ℓ_1, say. For $x = (\xi_1,\xi_2,\ldots)$, $y = (\eta_1,\eta_2,\ldots)$ define $T(x) = y$ by the rule $\eta_1 = \xi_2$, $\eta_{2n} = \xi_{2n+2}$, $n = 1,2,\ldots$, and $\eta_{2n+1} = \xi_{2n}$, $n = 1,2,\ldots$. That is

$$T(x) = (\xi_2, \xi_4, \xi_1, \xi_6, \xi_3, \xi_8, \ldots).$$

It is clear that T is a norm-preserving isomorphism of X onto X. In the above notation we define the operator U by the rule $U(x) = y$ where $\eta_1 = \xi_2$, $\eta_i = 0$, $i \neq 1$. Clearly $U \in K(X)$. Now consider $V = T - U$.

Let $x_n = \{\xi_i^n\}$ be defined by the rule

$$\xi_{2n}^n = 1, \quad \xi_i^n = 0, \quad i \neq 2n.$$

We show, by induction that, for each n, $V^n(x_n) = 0$ but $V^{n-1}(x_n) \neq 0$. The statement clearly holds for $n = 1$. Observe that for $m > 1$, $V(x_m) = x_{m-1}$. Then if the result holds for $m \geqq 1$, for $m + 1$ we have

$$V^{m+1}(x_{m+1}) = V^m V(x_{m+1}) = V^m(x_m) = 0,$$

$$V^m(x_{m+1}) = V^{m-1}V(x_{m+1}) = V^{m-1}(x_m) \neq 0.$$

This shows that V fails to have finite ascent. This example was taken from [68].

(1.4.6) COROLLARY (the Fredholm alternative).

If V is a Riesz-Schauder operator then V is one to one if and only if V is onto, i.e. $\alpha(V) = 0$ if and only if $\beta(V) = 0$.

Proof. Lemma 1.4.2 gives us that $a(V) = d(V) < \infty$. But if V is one to one, then $a(V) = 0$ so $d(V) = 0$, which means that V is onto. Therefore $\alpha(V) = \beta(V) = 0$.

Conversely if V is onto, then $d(V) = 0$ so $a(V) = 0$, which means V is one to one. Therefore $\alpha(V) = \beta(V) = 0$.■

(1.4.7) THEOREM

If $T \in K(X)$ and $\lambda \neq 0$ then $\alpha(\lambda I - T) = \beta(\lambda I - T) < \infty$.

Proof. By Theorem 1.4.5 (c), Theorem 1.2.1, and Proposition 1.2.7, we have that if we let $V = \lambda I - T$, then $\alpha(V) < \infty$ and $\beta(V) = \alpha(V^*) < \infty$.

We now prove that $\alpha(V) = \beta(V)$ by proving that $\alpha(V) = \alpha(V^*)$. Let $n = \alpha(V)$ and $m = \alpha(V^*)$. We first prove that $m \leq n$, for suppose $n < m$. Let $x_1, x_2, \ldots, x_n$ be a basis for $N(V)$ and $z_1^*, z_2^*, \ldots, z_m^*$ be a basis for $N(V^*)$. By elementary linear algebra there exist elements $x_1^*, x_2^*, \ldots, x_n^*$ in X and $z_1, z_2, \ldots, z_m$ in X such that $x_j^*(x_k) = \delta_{jk}$ for $j, k = 1, 2, \ldots, n$ and $z_j^*(z_k) = \delta_{jk}$ for $j, k = 1, 2, \ldots, m$.

Define the operator $U(x) = \sum_{j=1}^{m} x_j^*(x) z_j$ for all $x \in X$. Since $R(U)$ is finite dimensional it follows that $U \in K(X)$. It is easy to check that

$R(U) \cap R(V) = \{0\} = N(V) \cap N(U)$. We now examine the operator $W = V - U = \lambda I - (T + U)$. W is one to one since if $W(x) = 0$, then $V(x) = U(x)$ and so $x = 0$. Since $T + U \in K(X)$ we apply Corollary 1.4.6 to conclude that W is onto. Therefore there exists $y \in X$ such that $W(y) = z_{n+1}$. So $W^*(z^*_{n+1})(y) = z^*_{n+1}W(y) = z^*_{n+1}(z_{n+1}) = 1$. But a simple calculation shows that $z^*_{n+1}[R(W)] = \{0\}$. This contradiction proves that $m \leq n$, i.e. $\alpha(V^*) \leq \alpha(V)$. But since this applies to any Banach space X and any operator of the form $\lambda I - T$ with $T \in K(X)$ we apply the formula to V^{**} and V^* to get $\alpha(V^{**}) \leq \beta(V^*)$. This gives us $\alpha(V) = \alpha(V^{**}) \leq \alpha(V^*)$. So $\alpha(V) = \alpha(V^*)$.■

(1.4.8) THEOREM

If V_1 and V_2 are two permuting Riesz-Schauder operators, then V_1V_2 is also a Riesz-Schauder operator.

Proof. Let $W = V_1V_2$. That W is a Fredholm operator follows from Corollary 1.3.3. For a positive integer n, $W^n = V_1^n V_2^n$, where V_1^n, V_2^n and W^n are Fredholm operators and, in particular, have finite dimensional null-spaces. By linear algebra we see that

$$\alpha(W^n) = \alpha(V_2^n) + \dim[R(V_2^n) \cap N(V_1^n)].$$

But V_1 and V_2 have finite ascent and descent. Therefore $\alpha(W^n)$ has the same value for all $n \geq \delta$, where δ is the maximum of the ascents (and descents) of V_1 and V_2. Therefore $a(W) < \infty$. Also W^* is the product of two permuting Riesz-Schauder operators on X^*. Therefore $d(W) = a(W^*) < \infty$.■

By Corollary 1.4.6, any Riesz-Schauder operator V in $B(X)$ is one-to-one if and only if $R(V) = X$. Operators of the form $\lambda I - T$, where $\lambda \neq 0$ and $T \in K(X)$ are Riesz-Schauder operators by Theorem 1.4.5 and therefore have the property mentioned. As discussed in Chapter 2, the spectrum of T is the set of complex λ for which $\lambda I - T$ is not invertible in $B(X)$. It follows that the non-zero numbers λ in the spectrum of T must be eigenvalues. We conclude our introductory chapter with Riesz's theorem [55] on the eigenvalues of T.

(1.4.9) THEOREM

The set of eigenvalues of $T \in K(X)$ cannot have a non-zero point of accumulation.

Proof. Suppose otherwise. Then there exists a sequence $\{\lambda_j\}$ of distinct eigenvalues such that $\lambda_j \to \lambda \neq 0$. Choose, for each $j = 1,2,\ldots,$ an element x_j of X, $||x_j|| = 1$ where $T(x_j) = \lambda_j x_j$.

Consider the subsets $E_m = \{x_1,\ldots,x_m\}$ of X, $m = 1,2,\ldots$. We show by induction, that each set is a linearly independent set. This is obvious if $m = 1$. Suppose this holds for E_m. If E_{m+1} were not a set of linearly independent vectors, we could write

$$x_{m+1} = a_1x_1 + \cdots + a_mx_m \tag{1}$$

where the a_j are scalars. Applying the operator T we get

$$\lambda_{n+1}x_{m+1} = \lambda_1a_1x_1 + \cdots + \lambda_ma_mx_m. \tag{2}$$

Comparing (1) and (2) and using the fact that E_m is a linearly independent set, we see that $a_j = \lambda_ja_j/\lambda_{m+1}$ for $j = 1,\ldots,m$. As not all the $a_j = 0$, we get a contradiction.

Let L_m denote the linear span of E_m. Observe that $T(L_m) \subset L_m$. By Lemma 1.1.1 there exists, for $j = 2,3,\ldots,y_j \in L_j$, $||y_j|| = 1$ and $\text{dist}(y_j,L_{j-1}) \geqq \frac{1}{2}$. Note that if $y = a_1x_1 + \cdots + a_mx_m \in L_m$, then $T(y) - \lambda_my$ is a linear combination of $x_1,\ldots,x_{m-1}$ and is therefore in L_{m-1}.

Clearly $\{y_j/\lambda_j\}$ is a bounded sequence. Therefore $\{T(y_j)/\lambda_j\}$ contains a Cauchy subsequence. On the other hand let $p > q$ be two positive integers, then from the above

$$y_p - T(y_p)/\lambda_p \in L_{p-1}$$

and

$$T(y_q)/\lambda_q \in L_{p-1}$$

and therefore

$$||T(y_p)/\lambda_p - T(y_q)/\lambda_q|| \geqq \tfrac{1}{2}.$$

This contradiction completes our argument.■

We append here some historical remarks. After the pioneering efforts of Riesz and Schauder a breakthrough in the direction of the work of this monograph was made independently (and published in 1951) by three different authors, Atkinson [1], Gohberg [25], [26] and Yood [68]. Extensions of some of these results from the case of bounded linear operators to that of closed unbounded linear operators is important for applications and of considerable intrinsic interest. For thorough accounts see the survey article of Gohberg and Krein [27], the book of Kato [40] and the paper [62] of Schechter. However our program, which is to establish connections with Banach algebra theory, calls for a study of the bounded case only - the effort to treat the unbounded situation would, we feel, considerably complicate the exposition and divert attention from our objectives.

Chapter 2

BANACH ALGEBRAS

2.1 Introduction

Some of the basic facts and ideas of the theory of Banach algebras are essential to our development. These will be connected up to objects of great interest to us (Fredholm operators, etc.). The reader who has been exposed to the theory of Banach algebras should certainly skip ahead.

An algebra A over the complex (real) field which is a normed linear space in the norm $||x||$ is called a complex (real) normed algebra if $||xy|| \leq ||x||\ ||y||$ for all x,y in A. If A is complete in this norm it is called a Banach algebra. We shall confine our attention to the complex case, but point out that the theory of real normed Banach algebras can be reduced to the complex case [54].

For our later work the main examples of Banach algebras are derived from the algebra $B(X)$ of all bounded linear operators on the Banach space X with the usual linear operations, composition as multiplication and the operator norm. Of great importance for the development and applications of the subject are commutative Banach algebras of functions. The simplest non-trivial example is, perhaps, the algebra A of all continuous complex-valued functions on a locally compact Hausdorff space which 'vanish at infinity'. The algebraic operations are the pointwise operations and the norm is the sup norm.

As this last example emphasizes, a Banach algebra A need not have a multiplicative identity (which, if it exists, may be denoted by 1). If A has no identity, it can readily be isometrically embedded in a Banach algebra A_1 with identity 1 by defining A_1 to be the direct sum of A and the complex numbers and taking as norm $||x + \lambda|| = ||x|| + |\lambda|$, $x \in A$ and λ complex. (The multiplication in A_1 is given by $(x + \lambda)(y + \mu) = (xy + \lambda y + \mu x) + \lambda\mu$.) We refer to A_1 as the algebra obtained from A by adjoining an identity.

About thirty years ago Jacobson and others found the circle operation and the idea of quasi-inversion to be most helpful in a deeper study of rings and algebras without an identity.

The circle operation is introduced by the rule $x \circ y = x + y - xy$. One readily checks that it is an associative operation and that $0 \circ x = x \circ 0 = x$ for all $x \in A$. We say that $x \in A$ is right quasi-regular or r.q.r. (left quasi-regular or l.q.r.) if there exists $y \in A$ such that $x \circ y = 0$ $(y \circ x = 0)$. We call x quasi-regular (q.r.) if for some $y \in A$, $x \circ y = y \circ x = 0$. From the associativity of the circle operation it can be shown in a standard way that x is q.r. if and only if x is both l.q.r. and r.q.r. Moreover if there is a unique y such that $x \circ y = 0$ $(y \circ x = 0)$ then x is q.r. In particular if x is quasi-regular there is a unique y (denoted by x') such that $x \circ x' = x' \circ x = 0$ - all this calls attention to the following subsets of A. Let $Q_r (Q_\ell)$ denote the set of all r.q.r. (l.q.r.) elements of A and Q the set of q.r. elements. Then $Q = Q_r \cap Q_\ell$.

Of some importance for Banach algebra theory is the purely algebraic remark that $xy \in Q$ if and only if $yx \in Q$. The same is true if Q is replaced by Q_r or Q_ℓ. For let $yx \in Q_r$ with $(yx) \circ z = 0$. Then $(xy) \circ (xzy - xy) = x[(yx) \circ z]y = 0$.

(2.1.1) PROPOSITION

In a Banach algebra A, $x \in Q$ if $||x|| < 1$. The sets Q, Q_r and Q_ℓ are open.

Proof. Let $||x|| < 1$. It is clear that the infinite series

$$- \sum_{n=1}^{\infty} x^n$$

defines an element y in A. A simple computation shows that $x \circ y = y \circ x = 0$.

We show next that Q_r is open. Let $x \in Q_r$, $y \in A$ with $x \circ y = 0$. For $z \in A$ one has

$$z \circ y = (z - x + x) \circ y = (z - x) - (z - x)y$$

therefore $z \circ y \in Q$ if $||z - x|| < (1 + ||y||)^{-1}$. But then $z \circ [y \circ (z \circ y)'] = 0$ and $z \in Q_r$. Likewise Q_ℓ and therefore Q are open.■

(2.1.2) THEOREM

For a normed algebra A the mapping $x \to x'$ is a homeomorphism of Q onto Q.

Proof. Since $(x')' = x$ for $x \in Q$, it is enough to see that the mapping is continuous on Q. Let a be a fixed element of Q and let $b \in Q$ where, eventually, we let $b \to a$. First observe that

$$\begin{aligned} a' - b' &= a' \circ (b \circ b') - (a' \circ a) \circ b' \\ &= (a' \circ b) \circ b' - (a' \circ a) \circ b'. \end{aligned}$$

When this is expanded one readily obtains

$$a' - b' = (b - a) - a'(b - a) - (b - a)b' + a'(b - a)b'.$$

Now set $b - a = h$ and $b' - a' = k$. Our task is to show that $k \to 0$ if $h \to 0$. The last equations yield the inequality

$$||k|| \leqq ||h||(1 + ||a'|| + ||b'|| + ||a'|| \; ||b'||).$$

Inasmuch as $||b'|| \leqq ||a'|| + ||k||$ this gives

$$||k|| \leqq ||h||[(1 + ||a'||)^2 + ||k||(1 + ||a'||)].$$

Then

$$||k||[1 - ||h||(1 + ||a'||)] \leqq ||h||(1 + ||a'||)^2.$$

Therefore, provided that $||h|| < (1 + ||a'||)^{-1}$ we get

$$||k|| \leqq \frac{||h||(1 + ||a'||)^2}{1 - ||h||(1 + ||a'||)}.$$

This shows that $k \to 0$ as $h \to 0$.■

Suppose that A has an identity 1. Let $G_r(G_\ell)$ denote the set of right (left) regular elements (those with right (left) multiplicative inverses). Let $G = G_r \cap G_\ell$. The elements of G are called regular. The mapping $x \to 1 - x$ is a one-to-one mapping of Q_r onto G_r, of Q_ℓ onto G_ℓ and of Q onto G inasmuch as

$$(1 - x)(1 - y) = 1 - x \circ y.$$

From this we get the following.

(2.1.3) THEOREM

If A is a Banach algebra with identity 1, then G, G_ℓ and G_r are open. The mapping $x \to x^{-1}$ is a homeomorphism on G.

In particular G is a topological group (in the norm topology) which we examine briefly. Let G_0 be the connected component of G which contains 1. (G_0 is sometimes called the principal component of G).

(2.1.4) THEOREM

G_0 is an open and closed normal subgroup of G. The cosets of G_0 are the components of G. Furthermore G/G_0 is a discrete group.

Proof. G is an open subset of a Banach space. Therefore its components are open and closed subsets of G. For any $x \in G$ the mapping $y \to xy$ is a homeomorphism of G onto G. If, in addition, $x \in G_0$ then xG_0 is a component containing x. In other words, $xG_0 = G_0$ for each $x \in G_0$. Note that $x^{-1} G_0 \cup G_0$ is the union of two connected subsets of G each containing 1. Therefore $x^{-1} G_0 \cup G_0 = G_0$. It follows from all this that xy and x^{-1} are in G_0 if x and y lie in G_0. Thus G_0 is a subgroup of G. If $z \in G$ then $z^{-1} G_0 z = G_0$ or G_0 is a normal subgroup.

Observe that zG_0 is an open and closed connected subset of G for each $z \in G$. These sets are then components of G. If W is a component of G containing z, then $z^{-1} W = G_0$. Therefore every component has this form. Finally G/G_0 is a discrete group inasmuch as G_0 is open and closed in G.∎

2.2 On spectra

Let A be an algebra over the complex field with an identity 1. For $x \in A$ the spectrum of x, $sp(x|A)$, is defined to be the set of complex numbers λ for which $(\lambda - x)^{-1}$ does not exist in A. Note that if $\lambda \neq 0$, $(\lambda - x)^{-1}$ exists if and only if $\lambda^{-1}x$ is quasi-regular. This provides motivation for a definition of $sp(x|A)$, when A does not contain an identity, as the number 0 plus the set of all $\lambda \neq 0$ such that $\lambda^{-1}x$ is not quasi-regular. An advantage accrued by taking 0 in the spectrum of every $x \in A$ (when there is no identity) is that $sp(x|A) = sp(x|A_1)$, for all $x \in A$, where A_1 is the algebra obtained by adjoining an identity to A.

Let x be an element in a normed algebra A and λ, μ two scalars such that λx and μx are quasi-regular. Using the formula for $a' - b'$ developed in the proof of Theorem 2.1.2 we get

$$(\lambda - \mu)^{-1}[(\lambda x)' - (\mu x)'] = -x + (\lambda x)'x + x(\mu x)' - (\lambda x)'x(\mu x)'.$$

Now let D be the set of complex λ for which $(\lambda x)'$ exists. Our formula shows basically that $(\lambda x)'$ is a holomorphic A-valued function on the interior of D (int(D)). We can, however, avoid A-valued functions as follows. Let $x^* \varepsilon A^*$, the conjugate space of A. Form

$$\phi(\lambda) = x^*(\lambda x)'), \quad \lambda \varepsilon D.$$

Then, via Theorem 2.1.2, we see that $\phi(\lambda)$ is a holomorphic function of λ in int(D). In fact, for such λ,

$$\phi'(\lambda) = x^*(-x + (\lambda x)'x + x(\lambda x)' - (\lambda x)'x(\lambda x)').$$

(2.2.1) THEOREM

The spectrum of an element in a normed algebra A is non-void.

Proof. Let $x \varepsilon A$. We can suppose that A has an identity 1. If $sp(x|A)$ is void then $\lambda - x$ is regular for all complex λ. In particular $x \neq 0$. Now, for all $\lambda \neq 0$ we get

$$(1 - \lambda x)(1 - (\lambda x)') = (1 - (\lambda x)')(1 - \lambda x) = 1$$

which yields, after division by λ, the relation

$$1 - (\lambda x)' = \lambda^{-1}(\lambda^{-1} - x)^{-1}.$$

By Theorem 2.1.2, the right hand side approaches zero as λ approaches infinity. Thus

$$(\lambda x)' \to 1 \quad \text{as} \quad \lambda \to \infty.$$

From this, $||(\lambda x)'||$ is a bounded real-valued function of λ. On the other hand consider $x^* \varepsilon A^*$ and the corresponding function $\phi(\lambda) = x^*((\lambda x)')$. From our preliminary discussion $\phi(\lambda)$ is an entire function, and is bounded as $|\phi(\lambda)| \leqq ||x^*|| \; ||(\lambda x)'||$. By Liouville's theorem $\phi(\lambda)$ is a constant. Clearly $\phi(0) = 0$ so that $x^*(x') = 0$ for all $x^* \varepsilon A^*$ and therefore $x' = 0$. But then $x = 0$ which is impossible.■

An immediate consequence is the following important result of Gelfand and Mazur.

(2.2.2) THEOREM

A normed division algebra over the complex field is isomorphic to the complex number system.

Proof. Take $x \in A$. By Theorem 2.2.1 there exists a complex λ such that $\lambda - x$ is not regular. Then $x = \lambda$.■

(2.2.3) THEOREM

If A is a Banach algebra and $x \in A$, then $sp(x|A)$ is a compact set in the complex plane.

Proof. If $|a| > ||x||$ then, by Proposition 2.1.1, $a^{-1}x$ is quasi-regular and $a \notin sp(x|A)$. It follows from the same result that the complement of $sp(x|A)$ is open.■

This result fails to hold for normed algebras. If A is the algebra of all polynomials in a complex variable λ with complex coefficients (and the usual pointwise algebraic operations) and if one uses the norm

$$||f|| = \sup|f(\lambda)|, \quad |\lambda| \leq 1,$$

a normed algebra is formed where the spectrum of a non-constant f is the entire complex plane.

(2.2.4) LEMMA

Let A be an algebra over the complex field with identity. Let $p(\lambda)$ be a polynomial with complex coefficients. Then for each $x \in A$,

$$\{p(\lambda)|\lambda \in sp(x|A)\} = sp(p(x)|A).$$

Proof. Here, for $p(\lambda) = a_0 + a_1\lambda + \cdots + a_n\lambda^n$, $p(x)$ is defined to be the element in A given by $p(x) = a_0 + a_1x + \cdots + a_nx^n$.

For a complex number λ_0 consider the factorization

$$\lambda_0 - p(x) = a \prod_{j=1}^{n} (b_j - x)$$

where the b_j are the zeros of the polynomial $\lambda_0 - p(\lambda)$. We see that $(\lambda_0 - p(x))^{-1}$ fails to exist in A if and only if $(b_j - x)^{-1}$ fails to exist for some $j = 1,\ldots,n$. This shows that λ_0 is in $sp(p(x)|A)$ if and only if $\lambda_0 = p(\lambda)$ for some λ in $sp(x|A)$.■

If A has no identity the conclusion of Lemma 2.2.4 holds for all polynomials with zero constant term.

We define the spectral radius of x by

$$\rho(x|A) = \sup|\lambda|, \quad \lambda \in sp(x|A),$$

where there is no possible misunderstanding we replace $\rho(x|A)$ by $\rho(x)$ and $sp(x|A)$ by $sp(x)$. Lemma 2.2.4 assures us that $\rho(x^n) = [\rho(x)]^n$.

(2.2.5) THEOREM

Let x be an element in a Banach algebra A. Then the limit

$$\lim_{n\to\infty} ||x^n||^{1/n}$$

exists and equals $\rho(x)$.

Proof. By the proof of Theorem 2.2.3, $\rho(y) \leqq ||y||$ for all $y \in A$. Let $x \in A$. We then get

$$\rho(x) = [\rho(x^n)]^{1/n} \leqq ||x^n||^{1/n}$$

for all $n = 1,2,\ldots$. Consequently

$$\rho(x) \leqq \lim_{n\to\infty} \inf||x^n||^{1/n}. \tag{1}$$

Now take $x^* \in A^*$ and, as above, form $\phi(\lambda) = x^*[(\lambda x)']$. Note that λx is quasi-regular if $|\lambda| < ||x||^{-1}$ by Proposition 2.1.1, and furthermore

$$(\lambda x)' = -\sum_{n=1}^{\infty} \lambda^n x^n.$$

Then

$$\phi(\lambda) = -\sum_{n=1}^{\infty} \lambda^n x^*(x^n) = -\sum_{n=1}^{\infty} x^*(\lambda^n x^n) \tag{2}$$

for all $|\lambda| < ||x||^{-1}$. Note that $||x||^{-1} \leq \rho(x)^{-1}$ and that for any λ with $|\lambda| < \rho(x)^{-1}$, we have $|\lambda^{-1}| > \rho(x)$ and so $(\lambda^{-1})^{-1}x = \lambda x$ is quasi-regular. Therefore λx is quasi-regular if $|\lambda| < \rho(x)^{-1}$ and $\phi(\lambda)$ is analytic for such λ. Therefore the power series expansion (2) is valid for $|\lambda| < [\rho(x)]^{-1}$.

Let t be any positive number, $t > \rho(x)$. Then $t^{-1} < [\rho(x)]^{-1}$ and so

$$\phi(t^{-1}) = -\sum_{n=1}^{\infty} x^*(t^{-n}x^n). \tag{3}$$

Since (3) holds for all $x^* \in A^*$ we may apply the uniform boundedness theorem [22, p. 52] to assert that there is a number $M = M(t)$ such that

(4) $$||t^{-n}x^n|| \leq M, \quad n = 1,2,\ldots.$$

From (4) we obtain

$$||x^n||^{1/n} \leq M^{1/n}t.$$

This shows that

$$\limsup_{n\to\infty} ||x^n||^{1/n} \leq t.$$

However t is any number larger than $\rho(x)$. Therefore the lim sup does not exceed $\rho(x)$. In conjunction with (1) this completes the proof.■

An element x in a Banach algebra for which $\rho(x) = 0$ is called quasi-nilpotent.

2.3. Quotient algebra and ideals

(2.3.1) LEMMA

Let I be a closed two-sided ideal in a normed algebra A. Then A/I is a normed algebra in the quotient space norm.

Proof. The norm in A/I is given by $||x + I|| = \inf||x + y||$, $y \in I$. The multiplication in A/I is defined by

$$(x + I)(y + I) = xy + I.$$

Let $x_1 \in x + I$, $y_1 \in y + I$. Then $xy - x_1y_1 = (x - x_1)y + x_1(y - y_1) \in I$. Therefore $x_1y_1 \in xy + I$. Then $||xy + I|| = \inf\{||z||: z \in xy + I\}$ $\leq ||x_1y_1|| \leq ||x_1||\ ||y_1||$. Since $x_1(y_1)$ is an arbitrary element in $x + I(y + I)$ we see that $||xy + I|| \leq ||x + I||\ ||y + I||$.■

Of course A/I is a Banach algebra if A is a Banach algebra.

Recall from algebraic theory that a right (left) ideal I in A is called modular if there is a left (right) identity j for A modulo I, that is, if $jx - x \in I(xj - x \in I)$ for all $x \in A$. If $j \in I$, then $I = A$.

(2.3.2) PROPOSITION

In a Banach algebra A every modular maximal right (left) ideal is closed.

Proof. Let j be a left identity for A modulo a modular maximal right ideal M so that $jx - x \in M$ for all $x \in A$. We show that $||j - v|| \geqq 1$ for all $v \in M$.

Suppose otherwise, $||j - v|| < 1$ for some $v \in M$. By Proposition 2.1.1, $j - v$ is quasi-regular. Let w denote its quasi-inverse. Thus $(j - v) + w - (j - v)w = 0$. This gives

$$j = v + jw - w - vw.$$

But then $j \in M$ and $M = A$ which is impossible.

Since j is at a positive distance from M we see that the closure of M is also a modular right ideal distinct from A. By the maximality of M, M is closed.∎

For a more extended discussion of the Jacobson theory of rings and algebra than is appropriate here see [54, Chapter II] and the references cited there. Many equivalent formulations for the notion of the (Jacobson) radical $\mathcal{R}$ of a ring or algebra A are available. One procedure is to show that, given $x \in A$, xy is right quasi-regular for all $y \in A$ if and only if yx is left quasi-regular for all $y \in A$ and to define $\mathcal{R}$ as the set of all such x in A. $\mathcal{R}$ so defined can be shown to be the intersection of all the modular maximal right (left) ideals of A and is therefore a two-sided ideal of A. Still another approach can be made by representation theory [54].

The notion of a primitive ideal in A also has several equivalent formulations. A two-sided ideal I is called primitive if there is a modular maximal left ideal M such that $I = \{x \in A : xA \subset M\}$. An equivalent definition is employed in Chapter 5.

(2.3.3) PROPOSITION

In a Banach algebra A the radical and the primitive ideals are closed.

Proof. This follows from Proposition 2.3.3 and the above remarks.∎

As an aside we mention that the radical of A is the intersection of the primitive ideals of A.

It will be of importance to us whether or not A is <u>semisimple</u>, that is, the radical of A is $\{0\}$.

2.4. Topological divisors of zero

We now give a brief introduction to an idea which has considerable importance for the theory of Banach algebras, referring the reader to [54] for further results and for references to sources.

In a normed algebra A we call an element x a left (right) topological divisor of zero if there exists a sequence $\{x_n\}$ in A, each x_n of norm one, such that $xx_n \to 0(x_n x \to 0)$. A topological divisor of zero is an element which is either a left or right topological divisor of zero. We call x a two-sided topological divisor of zero if there is a sequence $\{x_n\}$ of elements of norm one such that both $xx_n \to 0$ and $x_n x \to 0$.

Consider the case where A has an identity. It is clear that if x is a left topological divisor of zero that x cannot have a left inverse.

(2.4.1) THEOREM

Let A be a Banach algebra with identity 1.

(a) If $x_n \in G_r(G_\ell)$ for $n = 1,2,...$ and $x_n \to x$ then either $x \in G_r(G_\ell)$ or x is a left (right) topological divisor of zero.

(b) If $x_n \in G$ for $n = 1,2,...$ and $x_n \to x$ then either $x \in G$ or x is a two-sided topological divisor of zero.

Proof. (a) Consider the case where each $x_n \in G_r$ and $x \notin G_r$. Consider a sequence $\{y_n\}$ where $x_n y_n = 1$, $n = 1,2,....$ We claim that the sequence $\{||y_n||\}$ is unbounded. For if $||y_n|| \leqq K$ for each n, then

$$1 - xy_n = (x_n - x)y_n \to 0.$$

Consequently $xy_n \in G$ for any n sufficiently large. This provides w_n where $xy_n w_n = 1$, contrary to $x \notin G_r$.

Without loss of generality we may (taking a subsequence if necessary) suppose that $||y_n|| \to \infty$. Set $z_n = ||y_n||^{-1} y_n$. We see that

$$xz_n = (x - x_n)z_n + ||y_n||^{-1} \to 0.$$

Then x is a left topological divisor of zero.

The proof of (b) follows along these lines.■

(2.4.2) COROLLARY

Let A be a Banach algebra with an identity. Then any x in the boundary of G is a two-sided topological divisor of zero.

Proof. Since G is open by Theorem 2.1.3, x cannot be in G. We apply Theorem 2.4.1.■

(2.4.3) THEOREM

In any normed algebra A the set of right (left, two-sided) topological divisors of zero is closed.

Proof. Let Z_r denote the set of right topological divisors of zero in A. Suppose $x_n \in Z_r$, $n = 1,2,\ldots$ and $x_n \to x$. For each positive integer m there exists $y_m \in A$ such that $||y_m|| = 1$ and $||y_m x_m|| < m^{-1}$. Turning to $y_m x$ we have $y_m x = y_m(x - x_m) + y_m x_m$ so that

$$||y_m x|| \leq ||x - x_m|| + m^{-1}.$$

Therefore $y_m x \to 0$ and $x \in Z_r$.■

2.5. On B(X) as a Banach algebra

Let X be a Banach space. We begin a consideration of $B(X)$ as a Banach algebra. This will lead to information on the Calkin algebra $B(X)/K(X)$. In this section we treat the notion of inversion. It is customary to use the symbol I for the identity operator.

By a bounded projection P of X onto the subspace E in X we mean an element $P \in B(X)$ where $P^2 = P$ and $R(P) = E$. It is readily seen that $R(P)$ must be closed. We need a well-known lemma due to F. J. Murray (1937).

(2.5.1) LEMMA

Let E be a closed linear manifold in the Banach space X. Then there exists a closed linear manifold F such that

$$X = E \oplus F$$

if and only if there is a bounded projection P of X onto E.

Proof. If the projection P exists, then $I - P$ is a bounded projection with closed range. Since we have the decomposition $x = P(x) + (I - P)(x)$ we see that

$$X = R(P) \oplus R(I - P).$$

Conversely suppose $X = E \oplus F$ where F is also a closed linear subspace. Each $x \in X$ has a unique representation $x = y + z$, $y \in E$, $z \in F$. Set $Q(x) = y$. Clearly $Q^2 = Q$ and $R(Q) = E$. We must show that Q is continuous. Let $x_n \to x$ in X and $Q(x_n) \to w$ in X. Since E is closed, $w \in E$. Now $x - Q(x_n) \in F$ and $x - Q(x_n) \to x - w$. Therefore, as F is closed, $x - w \in F$. Consequently we see that $w = Q(x)$. This makes Q into a closed linear mapping which, by the closed graph theorem [22, p. 57] must be continuous.■

(2.5.2) THEOREM

An element $T \in B(X)$ lies in G_ℓ if and only if T is one-to-one and there is a bounded projection P of X onto $R(T)$.

Proof. Suppose that $T \in G_\ell$. Then, for some $V \in B(X)$, $VT = I$. From this $(TV)^2 = TV$, so that TV is a bounded projection. Inasmuch as $(TV)T = T$ we get $R(T) = R(TV)$.

Conversely, suppose that T is one-to-one and the projection P exists. Then $R(T)$ is closed and T is a one-to-one continuous linear transformation of X onto $R(T)$. By [22, p. 57] the inverse mapping W is a continuous mapping of $R(T)$ onto X. We see that $WP \in B(X)$ and, moreover, $WPT = I$.■

(2.5.3) THEOREM

An element $T \in B(X)$ lies in G_r if and only if $R(T) = X$ and there exists a bounded projection of X onto $N(T)$.

Proof. Suppose that $T \in G_r$. Then, for some $U \in B(X)$, $TU = I$. From this we see directly that $R(T) = X$ and that $N(T) \cap R(U) = \{0\}$. By Theorem 2.5.2 we also know that $R(U)$ is closed.

Let $x \in X$. We can write $x = x - UT(x) + UT(x)$ where $x - UT(x) \in N(T)$ and $UT(x) \in R(U)$. This shows that

$$X = N(T) \oplus R(U).$$

By Lemma 2.5.1, there is a bounded projection of X onto $N(T)$.

Conversely, suppose that $R(T) = X$ and there is a bounded projection P of X onto $N(T)$. For each $x \in X$ we have the decomposition $x = P(x) + (I - P)(x)$ and see that $T(I - P)$ is a one-to-one continuous linear transformation of the Banach space $R(I - P)$ onto X. The inverse mapping W is a continuous linear mapping of X onto $R(I - P)$ by [22, p. 57]. Then $(I - P)W \in B(X)$ and furthermore $T(I - P)W = I$.∎

The following remark follows from Theorems 2.5.2 and 2.5.3 but is also immediate from first principles [22, p. 57].

(2.5.4) COROLLARY

Let $T \in B(X)$. Then $T \in G$ if and only if T is one-to-one and $R(T) = X$.

The notion of topological division of zero in $B(X)$ can also be characterized in terms of properties of linear operators.

(2.5.5) THEOREM

Let $T \in B(X)$. Then

(a) T is a left topological divisor of zero if and only if T is not one-to-one with closed range.

(b) T is a right topological divisor of zero if and only if $R(T) \neq X$.

Proof. (a) Suppose that T is a left topological divisor of zero in $B(X)$. There exists a sequence $\{V_n\}$ in $B(X)$ where each $||V_n|| = 1$ and $TV_n \to 0$. If T is one-to-one with closed range its inverse T^{-1} is continuous by [22, p. 57] and so $T^{-1}TV_n(x) \to 0$, uniformly for all x in X, $||x|| \leqq 1$. Then $||V_n|| \to 0$ which is impossible.

If T is not one-to-one with closed range there is [22, p. 57] a sequence $\{x_n\}$ in X, $||x_n|| = 1$, such that $T(x_n) \to 0$. For each n choose $x_n^* \in X^*$ with $||x_n^*|| = 1$. We define $U_n \in B(X)$ by the rule $U_n(x) = x_n^*(x)x_n$. Clearly $||U_n|| = 1$. Also $||TU_n(x)|| \leqq ||T(x_n)||$ for each x, $||x|| \leqq 1$. Therefore $TU_n \to 0$ in $B(X)$, so T is a left topological divisor of zero.

(b) Suppose that $R(T) = X$. If T is a right topological divisor of zero in $B(X)$ there is a sequence $\{U_n\}$ where each $||U_n|| = 1$ and $U_nT \to 0$. But then we have T^* as one-to-one on X with closed range, $||U_n^*|| = 1$ and $T^*U_n^* \to 0$. This is impossible by part (a) of this theorem.

Suppose that T is not a right topological divisor of zero in $B(X)$. We show that $R(T) = X$. Suppose otherwise. Then T^* does not have a continuous inverse and there exists a sequence $\{y_n^*\}$ in X^*, with every $||y_n^*|| = 1$ and $T^*(y_n^*) \to 0$. In other words $y_n^*T(x) \to 0$ uniformly for $||x|| \leq 1$. Let $w \in X$ have norm one and define U_n by the relation $U_n(y) = y_n^*(y)w$, $y \in X$, $n = 1,2,\ldots$. Clearly $||U_n|| = 1$. Also $U_nTx = y_n^*[T(x)]w \to 0$ uniformly for $||x|| \leq 1$. Then $U_nT \to 0$, which is impossible.■

The theory of §2.4 and §2.5 has some interesting consequences for $B(X)$.

(2.5.6) THEOREM

The following sets in $B(X)$ are open:

(a) The set of all $T \in B(X)$ which are one-to-one and have $R(T)$ closed.

(b) The subset of (a) for which $R(T) \neq X$.

(c) The set of all $T \in B(X)$ for which $R(T) = X$.

(d) The subset of (c) for which T is not one-to-one.

Proof. By Theorem 2.5.5, the complement of (a) is the set of left topological divisors of zero in $B(X)$, which is closed by Theorem 2.4.3. Likewise (c) is open.

Let T be in the set (b). As the set (a) is open there exists $\varepsilon_1 > 0$ so that the open ball in $B(X)$ with center at T and radius ε_1 lies in the set (a). We claim that for some ε_2, $0 < \varepsilon_2 < \varepsilon_1$, every $U \in B(X)$, $||U - T|| < \varepsilon_2$ has the property that $R(U) \neq X$. For otherwise we could find a sequence $\{U_n\}$, U_n one-to-one, $R(U_n) = X$ and $U_n \to T$. Each such U_n is regular (see Corollary 2.5.4). Applying Corollary 2.4.2 to this situation we see that T is a two-sided topological divisor of zero. This, however contradicts Theorem 2.5.5. Similar reasoning shows that the set (d) is open.■

In other words (such as will be employed later) the properties of (a), (b), (c), (d) are preserved under perturbation by bounded linear operators of sufficiently small norms.

An unsolved problem due to Banach [2] and going back at least to 1930 is the question of whether the set (b) must be non-void for every infinite-dimensional Banach space X.

Let $H_r(H_\ell)$ denote the complement in $B(X)$ of the set of all right (left) topological divisors of zero. These sets are open.

(2.5.7) COROLLARY

In $B(X)$, G is an open and closed subset of $H_r(H_\ell)$.

Proof. This is an immediate consequence of Theorems 2.1.3 and 2.5.6.∎

This reasoning also shows that G is an open and closed subset of $G_r(G_\ell)$ in $B(X)$.

We conclude this chapter with a few remarks about semisimplicity. Inasmuch as $B(X)$ is a primitive algebra, it is semisimple. On the other hand the Calkin algebra $C(X) = B(X)/K(X)$ may or may not be semisimple, depending on the choice of the Banach space X. In the classical case of a Hilbert space, $C(X)$ turns out to be a B*-algebra and hence semisimple (see [54, Chapter IV]). However consider $C(X)$ for the Banach space $X = L_1(\mu)$ of Lebesgue-integrable complex-valued functions $f(t)$ on the unit interval, that is for which

$$\int_0^1 |f|\, d\mu < \infty.$$

If T and U in $B(X)$ are weakly compact, then TU is compact as shown by Dunford and Pettis [21, p. 370]. From this it follows that the canonical image in $C(X)$ of the ideal of weakly compact operators on X is contained in the radical of $C(X)$. Hence $C(X)$ is not semisimple. These examples suggest the problem of determining criteria on X which would ensure the semisimplicity of $C(X)$.

Elements of $B(X)$ which under the natural homomorphism of $B(X)$ onto $C(X)$ map into the radical of $C(X)$ are called inessential and the set (ideal) of such elements denoted by $I(X)$. See [41] and [69].

Chapter 3

RIESZ OPERATORS

3.1 Introduction

Among the properties established for compact operators in Chapter 1 were the following: if T is a compact operator in $B(X)$, then for every complex non zero λ,

(3.1.1)

(i) $\lambda I - T$ has finite ascent and finite descent

(ii) $N[(\lambda I - T)^k]$ is finite dimensional for every $k = 1,2,3,\ldots$

(iii) $R[(\lambda I - T)^k]$ is closed with finite codimension for every $k = 1,2,3,\ldots$

(iv) the non zero points of $sp(T)$ consist of eigenvalues of T with zero as the only possible point of accumulation.

It is now natural to ask whether these properties characterize the class of compact operators. A moment's reflection is sufficient to convince oneself that this is not true, at least in the common infinite dimensional Banach spaces. For consider any operator T whose spectrum consists of the point zero alone (such operators are called quasinilpotent). Then evidently properties (3.1.1) are valid for such an operator. However, it is easy to find examples of noncompact quasinilpotent operators; we might take any infinite dimensional Banach space X and form a new Banach space X^2 whose elements consist of $X \times X$ with norm $||(x,y)|| = ||x|| + ||y||$. Then define $T \in B(X^2)$ by writing $T(x,y) = (y,0)$. It is easy to see that T is not compact but since $T^2 = 0$, we know that T is quasinilpotent. (It is perhaps worth remarking at this point that the existence of noncompact, quasinilpotent operators in a completely arbitrary Banach space seems to be an open question).

In view of the above situation, we will call any operator satisfying properties (3.1.1) a Riesz operator and we will write $R(X)$ for the class of Riesz operators in $B(X)$. It will turn out that $R(X)$ has many interesting and surprising properties; moreover, in some cases, it is easier to verify that an operator is in $R(X)$ than to show that it is compact. In questions of a spectral nature, it is often sufficient to know that the operator is in the larger class.

We now proceed to show that the class $R(X)$ of Riesz operators on a Banach space X can be characterized in various ways which are simpler than the defining properties (3.1.1).

3.2 Characterization in terms of the Fredholm region

Consider the following properties which are apparently very much weaker than (3.1.1):

For each $\lambda \neq 0$, suppose

(3.2.1)
(i) $N(\lambda I - T)$ is finite dimensional
(ii) $R(\lambda I - T)$ has finite codimension.

We will show that (3.2.1) is, in fact, sufficient to characterize $R(X)$.

Since the finite dimensional situation is obviously without interest in this context, we will assume once and for all that all Banach spaces mentioned in this chapter are infinite dimensional. We shall therefore obtain as the end result of this section the following theorem.

(3.2.2) THEOREM

An operator T in $B(X)$ is a Riesz operator if and only if, for each $\lambda \neq 0$, conditions (3.2.1) are valid.

Before proceeding to the proof of this theorem, it is convenient to express (3.2.1) in terms of the Fredholm region and to develop the basic properties of Fredholm operators.

(3.2.3) DEFINITION

If $T \in \Phi(X)$ (see Definition 1.3.1), then the integer $\alpha(T) - \beta(T)$ is called the index of T and will be denoted by $i(T)$.

For any operator T in $B(X)$, the Fredholm region is defined as $\{\lambda \in \mathbb{C} | \lambda I - T \in \Phi(X)\}$ and will be denoted by $\Phi(T)$. $\mathbb{C}$ denotes the set of complex numbers.

In these terms, Theorem 3.2.2 can be restated: An operator T in $B(X)$ is a Riesz operator if and only if $\Phi(T) = \mathbb{C} - \{0\}$. (During the development of the theory, we shall see what happens if $\Phi(T)$ is the entire complex plane.) In order to get the desired result, it is now necessary to develop a substantial part of the theory of Fredholm operators.

(3.2.4) LEMMA

Let X and Y be Banach spaces and T a bounded linear operator mapping X into Y. Suppose that N is a closed subspace such that $R(T) \oplus N$ is closed. Then $R(T)$ is closed.

Proof. Let T_o denote the operator defined on $X/N(T) \times N$ with range in Y given by $T_o([x],n) = Tx + n$ where $[x]$ is the coset of $X/N(T)$ containing x. Then it is easy to check that T_o is well defined, continuous, one-to-one and that its range is $R(T) \oplus N$. Since, by assumption $R(T) \oplus N$ is closed, T_o has a continuous inverse. Hence, there exists $k > 0$ such that

$$||Tx + n|| \geq k||([x],n)||$$

i.e.
$$||Tx + n|| \geq k(||[x]|| + ||n||).$$

In particular, when $n = 0$, we get

$$||Tx|| \geq k||[x]||$$

which implies that the map from $X/N(T)$ into Y has a continuous inverse. But this implies that it also has a closed range and this range is $R(T)$.■

(3.2.5) COROLLARY

If T is an operator in $B(X)$ with $X/R(T)$ finite dimensional, then $R(T)$ is closed.

Proof. $X = R(T) \oplus N$ for some finite dimensional N.■

(3.2.6) LEMMA

Let T be an operator in $B(X)$. Then T is Fredholm if and only if there exist bounded operators T_1 and T_2 and compact operators K_1 and K_2 such that

$$T_1T = I + K_1 \tag{1}$$

$$TT_2 = I + K_2 \tag{2}$$

Proof. Suppose $T \in \Phi(X)$. Then we can find closed subspaces X_1 and X_2 such that

$$X = N(T) \oplus X_1 = R(T) \oplus X_2.$$

Now T restricted to X_1 is an invertible operator; let $\hat{T}$ denote its inverse (defined on $R(T)$). Now we will define T_1 and T_2 to be the same operator: equal to $\hat{T}$ on $R(T)$ and equal to zero on X_2. Finally, let K_1 denote the operator which is $-I$ on $N(T)$ and zero on X_1, and K_2 the operator which is $-I$ on X_2 and zero on $R(T)$. It is then easy to verify that equations (1) and (2) are valid and obviously K_1 and K_2 are compact.

Conversely, if (1) and (2) are valid, then from (1), we see that $N(T) \subseteq N(I + K_1)$. Since K_1 is compact, we know that $N(I + K_1)$ is finite dimensional. Similarly, from (2), we see that $R(T) \supseteq R(I + K_2)$ and again, since K_2 is compact, $R(I + K_2)$ has finite codimension. Therefore, so does $R(T)$.■

COROLLARY

T_1 and T_2 can always be chosen to be Fredholm operators. For consider the definitions in the above proof.

(3.2.7) THEOREM

If T and S are Fredholm operators in $B(X)$, then so is TS. Moreover

$$i(TS) = i(T) + i(S).$$

Proof. Let $X_o = R(S) \cap N(T)$ and write

$$R(S) = X_o \oplus X_1$$
$$N(T) = X_o \oplus X_2$$
$$X = R(S) \oplus X_2 \oplus X_3.$$

Observe that we can always find such decompositions and X_2 and X_3 are finite dimensional. Then the result follows from the following equations which can be verified without great difficulty (although the first two require some thought for most people; complete details can be found in [29]):

$$\alpha(TS) = \alpha(S) + \dim X_o$$
$$\beta(TS) = \beta(T) + \dim X_3$$
$$\dim X_o + \dim X_2 = \alpha(T)$$
$$\dim X_2 + \dim X_3 = \beta(S).■$$

(3.2.8) THEOREM

An operator T in $B(X)$ is Fredholm if and only if $\pi(T)$ is invertible in the Calkin algebra $C(X)$.

Proof. If T is Fredholm, then by equations (1) and (2) of 3.2.6, we have

$$\pi(T_1)\pi(T) = \pi(I + K_1) = \pi(I)$$
$$\pi(T)\pi(T_2) = \pi(I + K_2) = \pi(I).$$

Thus $\pi(T)$ has a left and right inverse in $C(X)$. Therefore $\pi(T)$ is invertible. Conversely, if $\pi(T)$ has an inverse $\pi(S)$ in $C(X)$, then $\pi(TS - I) = \pi(ST - I) = 0$. Hence

$$TS = I + K_1$$

and

$$ST = I + K_2$$

for some compact operators K_1 and K_2. By 3.2.6, the result follows.■

(3.2.9) COROLLARY

If X is an infinite dimensional Banach space and $T \in B(X)$, then $\Phi(T)$ cannot be the entire complex plane.

Proof. If $\Phi(T) = \mathbb{C}$, then $\lambda I - T \in \Phi(X)$ for all λ. But by the above theorem, this implies that $\lambda\pi(I) - \pi(T)$ is invertible for all λ. Now since $\pi(I)$ is the identity element in $C(X)$, we would have $\pi(T)$ with empty spectrum. But this is impossible by Theorem 2.2.1.■

(3.2.10) THEOREM

If T is a Fredholm operator, then there exists $\varepsilon > 0$ such that, if $0 < |\lambda| < \varepsilon$, then $\lambda I - T$ is Fredholm and

(i) $\alpha(\lambda I - T)$ has constant value n with $n \leq \alpha(T)$

(ii) $\beta(\lambda I - T)$ has constant value m with $m \leq \beta(T)$

(iii) $i(\lambda I - T) = i(T)$.

Proof. Suppose T_1, T_2, K_1 and K_2 are the operators which satisfy equations (1) and (2) of 3.2.6. Then we have

$$T_1(\lambda I - T) = -K_1 - (I - \lambda T_1)$$
$$(\lambda I - T)T_2 = -K_2 - (I - \lambda T_2).$$

Choose $\varepsilon_1 = \max(||T_1||^{-1}, ||T_2||^{-1})$ so that for $|\lambda| < \varepsilon_1$, we know that $I - \lambda T_1$ and $I - \lambda T_2$ are invertible in $B(X)$. Hence the above equations can be written

$$-(I - \lambda T_1)^{-1} T_1 (\lambda I - T) = I + (I - \lambda T_1)^{-1} K_1$$

$$-(\lambda I - T) T_2 (I - \lambda T_2)^{-1} = I + K_2 (I - \lambda T_2)^{-1}.$$

Now 3.2.6 implies that $\lambda I - T$ is Fredholm. Applying the Index Theorem (3.2.7) to the first of these equations and recalling that $i(I + K) = 0$ for any compact K, we get

$$i(T_1) + i(\lambda I - T) = 0.$$

Similarly, from the second equation, we get

$$i(T_1) + i(T) = 0.$$

Hence assertion (iii) of the theorem is valid, for $\varepsilon \leq \varepsilon_1$.

Observe, now, that in the proof of Lemma 3.2.6, we chose T_1 such that $T_1 T x = x$ for $x \varepsilon X_1$. Hence $T_1(\lambda I - T)x = -(I - \lambda T_1)x$ for $x \varepsilon X_1$. This implies that $N[T_1(\lambda I - T)] \cap X_1 = \{0\}$ so that $N(\lambda I - T) \cap X_1 = \{0\}$. Hence

$$N(\lambda I - T) \oplus X_1 \subseteq X = N(T) \oplus X_1$$

so that $\alpha(\lambda I - T) \leq \alpha(T)$.

Similarly, using the fact that, if T is Fredholm, so is T^* and $\beta(T) = \alpha(T^*)$, we can deduce that $\beta(\lambda I - T) \leq \beta(T)$ for $|\lambda| < \varepsilon_2$ where ε_2 is obtained from T^* in the same way that ε_1 was obtained from T. If we take $\varepsilon_3 = \min(\varepsilon_1, \varepsilon_2)$, we see that $\alpha(\lambda I - T) \leq \alpha(T)$ and $\beta(\lambda I - T) \leq \beta(T)$ for $|\lambda| < \varepsilon_3$.

It remains to show that in a region $0 < |\lambda| < \varepsilon$, $\alpha(\lambda I - T)$ and $\beta(\lambda I - T)$ are constant in value.

Consider $x \varepsilon N(\lambda I - T)$. Since $Tx = \lambda x$, we get $T^2 x = \lambda T x = \lambda^2 x$, and generally, $T^n x = \lambda^n x$. Hence if $\lambda \neq 0$, we deduce that

$$N(\lambda I - T) \subseteq \bigcap_1^\infty R(T^n).$$

Now since T is Fredholm, so is T^n and from 3.2.5, $R(T^n)$ is a closed subspace for each n. Hence $\bigcap_1^{\infty} R(T^n)$ is a closed subspace which we will denote by $\tilde{X}$. Then obviously $T\tilde{X} \subseteq \tilde{X}$; in fact, since $R(T) \supseteq R(T^2) \supseteq \cdots$, it is easy to see that $T\tilde{X} = \tilde{X}$. Let $\tilde{T}$ denote the restriction of T to $\tilde{X}$. Clearly T is a Fredholm operator in $B(\tilde{X})$. By the previous part of the proof, we can find $\varepsilon_4 > 0$ such that, if $0 < |\lambda| < \varepsilon_4$, then $\lambda I - T_1$ is Fredholm with $i(\lambda I - T_1) = i(T_1)$, $\alpha(\lambda I - T_1) \leq \alpha(T_1)$ and $\beta(\lambda I - T_1) \leq \beta(T_1)$. But $\beta(T_1) = 0$ so $\beta(\lambda I - T_1) = 0$. Finally, since we showed earlier that $N(\lambda I - T) \subseteq \tilde{X}$, we can write

$$\alpha(\lambda I - T) = \alpha(\lambda I - T_1) = i(\lambda I - T_1) = i(T_1) = \alpha(T_1).$$

Since $\alpha(T_1)$ is independent of λ, we have shown that $\alpha(\lambda I - T)$ remains constant in $0 < |\lambda| < \varepsilon_4$, and since $i(\lambda I - T)$ remains constant, $\beta(\lambda I - T)$ must also remain constant in $0 < |\lambda| < \varepsilon_4$. Thus, taking $\varepsilon = \min(\varepsilon_3, \varepsilon_4)$, we get the required result.■

After these rather extensive preliminaries, we can begin the proof of the asserted characterization of $R(X)$. It should be noted, however, that the above properties of Fredholm operators have attracted much interest and they have been the object of extensive study. Some lines of development will be mentioned in §3.6. In later chapters, we will also need to obtain further properties of Fredholm operators and associated classes.

For any $T \in B(X)$ we define the <u>resolvent set of T</u> to be the set $res(T) = \mathbb{C} \backslash sp(T)$.

(3.2.11) THEOREM

If T is an operator in $B(X)$ with $\Phi(T) = \mathbb{C} - \{0\}$, then $sp(T)$ has no non zero points of accumulation.

Proof. For convenience, we will write $\alpha(\lambda)$ to denote $\alpha(\lambda I - T)$. Let $\lambda_o \in res(T)$ and suppose that, for some $\lambda_1 \neq 0$, $\alpha(\lambda_1) > 0$. Let Γ denote a curve joining λ_o to λ_1 without passing through the origin. For each $\mu \in \Gamma$, $\mu I - T$ is a Fredholm operator. By 3.2.10, there exists $\varepsilon = \varepsilon(\mu) > 0$ such that $\alpha(\lambda) \leq \alpha(\mu)$ in the open disc $S(\mu)$ with center μ and radius ε; moreover $\alpha(\lambda)$ is constant in $0 < |\lambda - \mu| < \varepsilon$. Since Γ is compact, we can obtain a finite set $\{\mu_1, \mu_2, \ldots, \mu_n\}$ of points on Γ such that $\mu_1 = \lambda_o$, $\mu_n = \lambda_1$ and $\Gamma \subset \bigcup_1^n S(\mu_i)$ with $S(\mu_i) \cap S(\mu_{i+1}) \neq \emptyset$

for $i = 1,2,\ldots,n-1$. Now since $\alpha(\lambda_o) = 0$, it follows that $\alpha(\lambda) = 0$ for all λ in $S(\mu_1)$. Hence, because $S(\mu_2)$ overlaps $S(\mu_1)$, the constant value of $\alpha(\lambda)$ in the region $0 < |\lambda - \mu_2| < \varepsilon(\mu_2)$ must be zero. By proceeding along the family of discs, we finally deduce that $\alpha(\lambda) = 0$ in the region $0 < |\lambda - \lambda_1| < \varepsilon(\lambda_1)$.

By using the same argument for $\beta(\lambda)$, we see that both $\alpha(\lambda)$ and $\beta(\lambda)$ must be zero in the neighbourhood of λ_1, i.e. λ_1 is an isolated spectral point.

Finally, suppose $\beta(\lambda_1) > 0$ for some $\lambda_1 \neq 0$. Making use, once again, of the fact that $\beta(\lambda I - T) = \alpha(\bar{\lambda} I - T^*)$, we can obtain the same conclusion as before. Since any point λ_1 in sp(T) must have either $\alpha(\lambda_1)$ or $\beta(\lambda_1)$ different from zero, we have the result.■

We are now ready for the long-heralded result:

(3.2.12) THEOREM

If T is an operator in B(X) with $\Phi(T) = \mathbb{C} - \{0\}$, then T is a Riesz operator.

Proof. We need to verify properties (i) - (iv) of 3.1.1. If T is quasinilpotent, there is nothing to prove, so assume sp(T) contains a non zero point λ_1. By 3.2.11, we know that λ_1 is isolated and hence, using the operational calculus

$$f \to \frac{1}{2\pi i} \int_c f(\lambda)(\lambda I - T)^{-1} d\lambda$$

we can obtain a projection P_1 associated with the spectral set $\{\lambda_1\}$. Let X_1 denote the range of P_1 and X_2 its nullspace. Then $X = X_1 \oplus X_2$ and X_1 and X_2 are invariant under T. Let T_i denote the restriction of T to X_i, $i = 1,2$. Then we know from standard theory that $sp(T_1) = \{\lambda_1\}$. Moreover, obviously $\alpha(\lambda_1 I - T_1) < \infty$. We will now show that $\beta(\lambda_1 I - T_1) < \infty$. Since $sp(T_2) = sp(T) - \{\lambda_1\}$, we can write

$$\begin{aligned} R(\lambda_1 I - T) &= (\lambda_1 I - T)X_1 \oplus (\lambda_1 I - T)X_2 \\ &= R(\lambda_1 I - T_1) \oplus X_2. \end{aligned}$$

Suppose $X_1 = R(\lambda_1 I - T) \oplus Y_1$; then

$$X = R(\lambda_1 I - T_1) \oplus Y_1 \oplus X_2$$

$$= R(\lambda_1 I - T) \oplus Y_1.$$

Hence Y_1 is finite dimensional and, in fact, $\beta(\lambda_1 I - T_1) = \beta(\lambda_1 I - T) < \infty$. Thus the Fredholm set of T_1 is the entire complex plane so that, by 3.2.9, X_1 must be finite dimensional. We now proceed to verify the required properties.

Firstly, since $\lambda_1 \in res(T_2)$, we know that $N[(\lambda_1 I - T)^n] = N[\lambda_1 I - T_1)^n]$ and since T_1 acts in a finite dimensional space, the ascent of $\lambda_1 I - T_1$ must be finite and each space $N[(\lambda_1 I - T)^n]$ must be finite dimensional.

Similarly, $R[(\lambda_1 I - T)^n] = R[(\lambda_1 I - T_1)^n] \oplus X_2$ and, for the same reasons, the required descent properties are valid. Moreover, if $X_1 = R[(\lambda_1 I - T_1)^n] \oplus Y_n$, then $X = R[(\lambda_1 I - T_1)^n] \oplus Y_n \oplus X_2 =$ $= R[(\lambda_1 I - T)^n] \oplus Y_n$ so that $R[(\lambda_1 I - T)^n]$ has finite codimension. Hence properties (i) - (iii) of 3.1.1 are verified. Finally, we observe that λ_1 must be an eigenvalue since otherwise $\lambda_1 I - T_1$ would map X_1 onto itself and so $sp(T_1)$ would be empty. This completes the proof.■

3.3 Characterization in terms of the Calkin algebra

It is now a simple matter to show the following result:

(3.3.1) THEOREM

An operator T in $B(X)$ is a Riesz operator if and only if

$$\lim_{n\to\infty} \inf_{K\in K(X)} ||T^n - K||^{1/n} = 0.$$

Proof. By Theorem 3.2.12, we know that T is a Riesz operator if and only if $\lambda I - T$ is Fredholm for every $\lambda \neq 0$. By Theorem 3.2.8, we know that this is equivalent to saying that $\lambda\pi(I) - \pi(T)$ is invertible in the Calkin algebra $C(X)$, i.e. that $\pi(T)$ is quasinilpotent in $C(X)$. But by Theorem 2.2.5, this is the same as the condition

$$\lim_{n\to\infty} ||\pi(T^n)||^{1/n} = 0$$

and since $||\pi(T^n)|| = \inf_{K\varepsilon K(X)} ||T^n - K||$, the result follows.■

3.4 Characterizations in terms of the resolvent operator

(3.4.1) THEOREM

Let T be an operator in $B(X)$. Then T is a Riesz operator if and only if its resolvent $R_\lambda(T)$ can be written

$$R_\lambda(T) = C(\lambda) + B(\lambda) \quad \text{for all} \quad \lambda \varepsilon \text{res}(T)$$

where $C(\lambda)$ is compact and $B(.)$ is analytic for all non zero λ. Here $R_\lambda(T) \equiv (\lambda I - T)^{-1}$.

Proof. Suppose the resolvent of T satisfies the above condition. Then, for $|\lambda| > \rho(T)$, we have power series expansions for $R_\lambda(T)$ and $B(\lambda)$ in powers of $\frac{1}{\lambda}$. Hence $C(\lambda)$ has a similar expansion. Write

$$\sum_0^\infty \lambda^{-n-1} T^n = \sum_0^\infty \lambda^{-n-1} C_n + \sum_0^\infty \lambda^{-n-1} B_n.$$

Since $B(\frac{1}{\lambda})$ is entire, we know that $||B_n||^{1/n} \to 0$. Moreover, each C_n is compact since we can calculate C_n by taking the appropriate derivative of $C(\lambda)$ and such derivatives must be compact since $K(X)$ is a closed ideal. Finally, we can also see that $T^n = C_n + B_n$ so that $||B_n||^{1/n} \to 0$ gives $||T^n - C_n||^{1/n} \to 0$. By Theorem 3.3.1, the result follows.

Conversely, if T is a Riesz operator, then there exist compact operators C_n such that $||T_n - C_n||^{1/n} \to 0$. Let $B_n = T^n - C_n$. Now consider $|\lambda| > \rho(T)$. Then

$$R_\lambda(T) = \sum_0^\infty \lambda^{-n-1} T_n = \sum_0^\infty \lambda^{-n-1} C_n + \sum_0^\infty \lambda^{-n-1} B_n.$$

We know that the series $\sum_0^\infty \lambda^{-n-1} T^n$ converges for $|\lambda| > \rho(T)$ and the series $\sum_0^\infty \lambda^{-n-1} B_n$ converges for all non zero λ. Hence $\sum_0^\infty \lambda^{-n-1} C_n$ must converge at least for $|\lambda| > \rho(T)$. Call its sum function $C(\lambda)$. Then $\pi C(\lambda) = 0$ on the open set $|\lambda| > \rho(T)$. Obviously $C(\lambda)$ has an analytic continuation to all of $\mathrm{res}(T)$ which can be defined by writing

$$\tilde{C}(\lambda) = R_\lambda(T) - \sum_0^\infty \lambda^{-n-1} B_n.$$

Hence $\pi\tilde{C}(\lambda)$ is an analytic continuation of $\pi C(\lambda)$ and since $\mathrm{res}(T)$ is connected, we must have $\pi\tilde{C}(\lambda) = 0$ for all $\lambda \in \mathrm{res}(T)$. Thus $\tilde{C}(\lambda)$ is compact for all $\lambda \in \mathrm{res}(T)$. This completes the proof.■

We know that the resolvent operator $R_\lambda(T) = (\lambda I - T)^{-1}$ is analytic on the resolvent set. Hence an isolated spectral point can be thought of as an isolated singularity of $R_\lambda(T)$. The usual classification in terms of the Laurent series is then available and we will be concerned with the case where the isolated spectral point is a pole of the resolvent operator. If the associated spectral projection (which is also the residue of $R_\lambda(T)$ at the point in question) is finite dimensional, then we speak of a <u>pole of finite multiplicity</u>; the dimension of the range of the projection is referred to as the <u>multiplicity</u> of the point.

In this section, we will show that an operator is a Riesz operator if and only if each non zero point of its spectrum is a pole of finite multiplicity. In order to do this, we will prove a comprehensive lemma which will reveal the role of ascent and descent in a new light.

First, however, we recall a result from Chapter 1: if T is an operator in $B(X)$ and $a(T)$ and $d(T)$ are finite, then they are equal.

Now we can state the basic decomposition lemma.

(3.4.2) LEMMA

Suppose T is an operator in $B(X)$ and $a(T)$ and $d(T)$ are finite with common value $p \neq 0$. Then

(i) X can be written

$$X = R(T^p) \oplus N(T^p)$$

(ii) The subspaces $R(T^p)$ and $N(T^p)$ are both closed and invariant under T (i.e. the decomposition <u>completely reduces</u> T.)

(iii) T maps $R(T^p)$ onto itself in a one-to-one manner and T restricted to $N(T^p)$ is nilpotent

(iv) $\lambda = 0$ is an isolated point of $sp(T)$

(v) $\lambda = 0$ is a pole of $R_\lambda(T)$ of order p

(vi) The spectral projection corresponding to $\lambda = 0$ has $N(T^p)$ as its range and $R(T^p)$ as its nullspace.

Conversely, if $\lambda = 0$ is a pole of $R_\lambda(T)$ of order p, then $a(T) = d(T) = p$.

Proof. (i) If $x \in R(T^p) \cap N(T^p)$, then $x = T^p y$ for some $y \in X$ and also $T^p x = 0$. Hence $T^{2p} y = 0$ so that $y \in N(T^{2p}) = N(T^p)$, i.e. $T^p y = 0 = x$. Hence $R(T^p) \cap N(T^p) = \{0\}$. Now it is clear that $R(T^p)$ and $N(T^p)$ are invariant under T. Let T_1 denote the restriction of T to $R(T^p)$. Then T_1 maps $R(T^p)$ onto itself; in fact, $R(T_1^p) = R(T^p)$. Hence, given $x \in X$, there exists $x_1 \in R(T^p)$ such that $T^p x = T_1^p x_1$. Rewrite this as $T^p x = T^p x_1$ and define $x_2 = x - x_1$ so that $x_2 \in N(T^p)$. Hence the decomposition asserted in (i) is verified.

(ii) Obviously $N(T^p)$ is closed and (i), together with Lemma 3.2.4 imply that $R(T^p)$ is closed.

(iii) We have already observed that T maps $R(T^p)$ onto itself; to show that this is a one-to-one map, suppose $T_1 x = 0$ for some $x \in R(T^p)$. Then we would have

$$x \in N(T_1) \cap R(T^p) \subseteq N(T) \cap R(T^p) \subseteq N(T^p) \cap R(T^p) = \{0\}.$$

Hence T_1 is one-to-one.

Next, we observe that if $x \in N(T^p)$, then $T^p x = 0$ so that T restricted to $N(T^p)$ is nilpotent.

(iv) We know that $0 \in res(T_1)$. Since $res(T_1)$ is an open set, there exists $\varepsilon > 0$ such that $|\lambda| < \varepsilon$ implies that $\lambda \in res(T_1)$ and hence that $\alpha(\lambda I - T_1) = \beta(\lambda I - T_1) = 0$. But if $\lambda \neq 0$, we know from the proof of Theorem 3.2.10, that $N(\lambda I - T) \subseteq \bigcap_1^\infty R(T^p)$ so that we have $N(\lambda I - T) \subseteq N(\lambda I - T_1) = \{0\}$. Hence $\alpha(\lambda I - T) = 0$ for $0 < |\lambda| < \varepsilon$. By repeating all the above arguments from the beginning of the proof with T^* in place of T we come to a similar conclusion for $\alpha(\lambda I - T^*)$, i.e. about $\beta(\bar{\lambda}I - T)$. Thus, for a suitably small ε, $0 < |\lambda| < \varepsilon$ implies $\alpha(\lambda I - T) = \beta(\lambda I - T) = 0$, i.e. $\lambda = 0$ is an isolated point of $sp(T)$.

(v) Now write $X_1 = R(T^p)$ and $X_2 = N(T^p)$. Then the decomposition $X = X_1 \oplus X_2$ completely reduces $(\lambda I - T)^{-1}$ for every λ in res T. If we write T_i for the restrictions of T to X_i, $(i = 1,2)$, then the restriction of $(\lambda I - T)^{-1}$ to X_i is $(\lambda I - T_i)^{-1}$. Hence, for any function appropriate to the operational calculus, we see that, since

$$f(T) = \frac{1}{2\pi i} \int_c f(\lambda)\, R_\lambda(T) d\lambda$$

then $f(T)$ is completely reduced by the decomposition $X_1 \oplus X_2$.

Now suppose $R_\lambda(T)$ has a Laurent decomposition

$$R_\lambda(T) = \sum_{-\infty}^{\infty} B_n \lambda^n.$$

It is well known (see [64, p. 305]) that the formulas for coefficients of Laurent series in the conventional sense have a natural generalization to the operator-valued case. In particular, each B_n is given by $f_n(T)$ for certain appropriate functions f_n. Hence all B_n are completely reduced by $X_1 \oplus X_2$. Let $B_n^{(i)}$, $(i = 1,2)$, denote the restrictions of B_n to X_i. Then it is not surprising and certainly not difficult to verify that

$$(\lambda I - T_i)^{-1} = \sum_{-\infty}^{\infty} B_n^{(i)} \lambda^n.$$

Now since $0 \in res(T_1)$, we know that $B_{-n}^{(1)} = 0$ for $n = 1,2,3,\ldots$. Also since T_2 is nilpotent, we know that the spectral radius of T_2 is zero and hence $(\lambda I - T_2)^{-1}$ has a Neumann series which terminates after p terms:

$$(\lambda I - T_2)^{-1} = \sum_{1}^{p} \lambda^{-n} T_2^{n-1}.$$

By the uniqueness of series representations, we know that $B_{-n}^{(2)} = T_2^{n-1}$ for $1 < n \leq p$ and $B_{-n}^{(2)} = 0$ for $n > p$.

Now, if P is the projection onto $N(T^p)$ parallel to $R(T^p)$, we know that for $n = 1,2,3,\ldots,$

$$\begin{aligned} B_{-n} &= B_{-n}(I - P) + B_{-n}P \\ &= B_{-n}^{(1)}(I - P) + B_{-n}^{(2)}P \\ &= 0 + B_{-n}^{(2)}P \ . \end{aligned}$$

Hence $B_{-n} = 0$ for $n > p$ and $B_{-p} \neq 0$ so that $\lambda = 0$ is a pole of order p.

(vi) We know that the spectral projection corresponding to $\lambda = 0$ is the operator B_{-1}. Hence $B_{-1} = B_{-1}^{(2)}P = T_2^oP = P$.

It now remains to prove the converse statement. Let $\lambda = 0$ be a pole of $R_\lambda(T)$ of order p. Write P_o for the projection associated with $\{\lambda_o\}$ and T. Then the formulas for the principal part of the Laurent series for $R_\lambda(T)$ in powers of λ give

$$T^{p-1}\ P_o \neq 0 = T^p\ P_o .$$

Hence $R(P_o) \subseteq N(T^p)$. The reverse inclusion is valid also as can be seen by recalling the well known formula [15, p. 309] :

$$R(P_o) = \{x \in X : ||T^n x||^{1/n} \to 0\}.$$

Thus $R(P_o) = N(T^p)$.

Suppose now that $x \in N(T^{p+k})$; then clearly $x \in R(P_o)$ so that $N(T^{p+k}) = N(T^p)$ for each positive integer k. Thus $a(T) \leq p$. But if $N(T^{p-1}) = N(T^p) = R(P_o)$, we can deduce $T^{p-1}P_o = 0$ which we know is false. Hence $a(T) = p$.

Next, observe that since P_o commutes with T, we have

$$P_o\ T^{p-1} \neq 0 = P_o\ T^p .$$

Thus $R(T^p) \subseteq N(P_o)$. Let T_1 denote the restriction of T to $N(P_o)$; it is then well known that $0 \in res(T_1)$ so that $R(T_1^k) = N(P_o)$ for all k. Thus $R(T^k) \supseteq N(P_o)$ so that we have the equality $R(T^p) = N(P_o)$. Now if

$R(T^{p-1}) = R(T^p) = N(P_o)$, then we get $P_o T^{p-1} = 0$, again contrary to a previous equation. Thus $d(T) \geq p$. Also, since $R(T^k) \supseteq N(P_o)$ for all k, we see that $R(T^{p+1}) = N(P_o)$ so that $d(T) \geq p$. Hence $a(T) = d(T) = p$.■

(3.4.3) THEOREM

An operator T in $B(X)$ is a Riesz operator if and only if the non zero points of $sp(T)$ are poles of finite multiplicity.

Proof. If T is a Riesz operator, then by the above decomposition lemma, the non zero points of $sp(T)$ are poles. Moreover, since $N[(\lambda_o I - T)^p]$ is the range of the spectral projection corresponding to $\{\lambda_o\}$ and, by assumption, this is a finite dimensional space, each pole has finite multiplicity.

Conversely if λ_o is a pole of finite multiplicity, we know from Lemma 3.4.2 that $\lambda_o I - T$ has finite ascent and descent. Let p denote the common value. Then $N[(\lambda_o I - T)^p] = R(P_o)$ and since the latter is finite dimensional, $\alpha(\lambda_o I - T) < \infty$. Similarly, $R[(\lambda_o I - T)^p] = N(P_o)$ and hence $R[(\lambda_o I - T)^p]$ has finite codimension. But $R(\lambda_o I - T) \supseteq R[(\lambda_o I - T)^p]$ so that $\beta(\lambda_o I - T) < \infty$. Thus we can deduce that $\Phi(T) = \mathbb{C} - \{0\}$ and hence T is a Riesz operator.■

COROLLARY

An operator T in $B(X)$ is a Riesz operator if and only if the non zero points of $sp(T)$ are isolated and the corresponding spectral projections are all finite dimensional.

Proof. If T is a Riesz operator, then the conclusion is obvious from Theorem 3.4.3. Conversely, if λ_o is a non zero point of $sp(T)$ and the corresponding spectral projection P_o is finite dimensional, we can proceed very much as in the proof of Lemma 3.4.2 (v). Write $X_1 = R(P_o)$ and $X_2 = N(P_o)$, let T_i denote the restriction of T to X_i, $i = 1,2$, and consider the Laurent expansion in powers of $\lambda - \lambda_o$. Then if $\{B_n\}_{-\infty}^{\infty}$ are the Laurent coefficients, we have

$$(\lambda I - T_i)^{-1} = \sum_{-\infty}^{\infty} B_n^{(i)} (\lambda - \lambda_o)^n \quad (i = 1,2).$$

But X_1 is finite dimensional and $\lambda_o \in sp(T_1)$ so that λ_o must be a pole of $(\lambda I - T_1)^{-1}$ since in a finite dimensional space, the spectrum of an operator consists entirely of poles of the resolvent. On the other hand $\lambda_o \in \text{res}(T_2)$ so that $(\lambda I - T_2)^{-1}$ is analytic at λ_o. Since $(\lambda I - T)^{-1}$ is the direct sum of the two resolvents $(\lambda I - T_i)^{-1}$ we see that λ_o must be a pole of $(\lambda I - T)^{-1}$. Theorem 3.4.3 then implies that T is a Riesz operator.■

3.5 The West decomposition

If C is a compact operator and Q is quasinilpotent, then clearly $C + Q$ is a Riesz operator. The converse question is much less trivial, i.e. can every Riesz operator be written as a sum, $C + Q$. In the case where the operators are defined on a Hilbert space, T. T. West [66] was able to obtain an affirmative result; the general problem is still unsettled and constitutes an important open question in this area. Even a simplified version of West's proof would be a considerable step forward.

Before turning to West's proof, it is interesting to note that the above decomposition problem has a simple interpretation in terms of the Calkin algebra. In fact, the problem has an affirmative solution if and only if π maps the class of quasinilpotent operators in $B(X)$ onto the class of quasinilpotent elements in $C(X)$.

(3.5.1) LEMMA

Let T be a Riesz operator on a Banach space X and let M be a closed invariant subspace for T. Then, if T_M denotes the restriction of T to M,

(i) T_M is a Riesz operator

(ii) $sp(T_M) \subseteq sp(T)$.

Proof. We begin by showing that for any λ in res T, $(\lambda I - T)M = M$. Suppose that $|\lambda| > \rho(T)$ so that $R_\lambda(T) = (\lambda I - T)^{-1} = \sum_0^\infty \lambda^{-n-1} T^n$. Clearly $R_\lambda(T)M \subseteq M$. Then choose x in M and x^* in $M^\perp$, the annihilator of M (i.e. such that $x^*(M) = 0$). Since $x^*R_\lambda(T)x = 0$ for all $|\lambda| > \rho(T)$ and since $\text{res}(T)$ is connected, we can deduce by analytic continuation that $x^*R_\lambda(T) = 0$ for all $\lambda \in \text{res}(T)$. Since this is valid for each x in M and each x^* in $M^\perp$, we deduce that $R_\lambda(T)M \subseteq M$ for any $\lambda \in \text{res}(T)$. This, along with the obvious inclusion $(\lambda I - T)M \subseteq M$, implies that $(\lambda I - T)M = M$.

From this, it quickly follows that $sp(T_M) \subseteq sp(T)$. Also it is clear that $(\lambda I - T_M)^{-1} = (\lambda I - T)^{-1}|M$ for all $\lambda \in res(T)$. Hence if λ_o is a non zero point in $sp(T_M)$ and P_M is the associated spectral projection in $B(M)$, we have

$$P_M = \frac{1}{2\pi i}\int_c (\lambda I - T_M)^{-1} d\lambda = \frac{1}{2\pi i}\int_c [(\lambda I - T)^{-1}|M] d\lambda$$

$$= [\frac{1}{2\lambda i}\int_c (\lambda I - T)^{-1} d\lambda]|M$$

$$= P_o|M$$

where c is a suitable curve separating λ_o from the remainder of $sp(T)$ and P_o is the spectral projection associated with λ_o and T. Now since T is a Riesz operator, we know that P_o is finite dimensional. Hence so is P_M. By the corollary to Theorem 3.4.3, this implies that T_M is a Riesz operator. The proof of the lemma is complete.■

REMARK

Notice that the first part of the argument depends only on the fact that T has a connected resolvent set. For resolvent sets in general, we can only deduce that $sp(T_M)$ is contained in "the spectrum of T with its holes filled in", i.e. $sp(T_M)$ lies in the complement of the unbounded component of $res(T)$. This fact is proved in [14].

(3.5.2) THEOREM

If X is a Hilbert space and T is a Riesz operator on X, then T can be written as $C + Q$ where C is compact and Q is quasinilpotent.

Proof. If $sp(T)$ consists only of $\lambda = 0$, then the theorem is trivial. If $sp(T)$ is a finite set then again we can obtain the result without difficulty by the use of spectral projections viz. $\lambda = 0$ is an isolated point of $sp(T)$ in this case so there is a corresponding spectral projection P; then $T = TP + T(I - P)$ with TP quasinilpotent and $T(I - P)$ finite dimensional. We shall therefore assume that $\{\lambda_1, \lambda_2, \ldots\}$ is an enumeration of the non zero points of $sp(T)$ with $|\lambda_1| \geq |\lambda_2| \geq \cdots$. Let P_j denote the spectral projection associated with the point $\{\lambda_j\}$ and define

$$L_K = \bigoplus_{1}^{K} R(P_j)$$

so that $\{L_K\}$ forms an increasing sequence of finite dimensional subspaces, each invariant under T. We can therefore interpolate as follows: there exists a sequence $\{M_j\}$ of subspaces of X such that each M_j is invariant under T, $M_j \subseteq M_{j+1}$, $\dim M_j = j$ and $\{L_K\}$ is a subsequence of $\{M_j\}$. The existence of such a sequence is a consequence of well known facts about operators on finite dimensional spaces. Now consider an orthonormal system $\{e_j\}$ where $e_j \in M_j \backslash M_{j-1}$. Then there exists $\alpha_j \in \mathbb{C}$ and $f_{j-1} \in M_{j-1}$ such that

$$Te_j = \alpha_j e_j + f_{j-1}.$$

Elementary matrix considerations make it clear that $\alpha_j \in sp(T)$ and $|\alpha_j| \geq |\alpha_{j+1}|$ for all j.

Now every x in X can be written as

$$x = \sum_j (x,e_j)e_j + y$$

where $y \perp e_j$ for all j. Hence

$$Tx = \sum_j \alpha_j (x,e_j)e_j + \sum_j (x,e_j)f_{j-1} + Ty.$$

We now define operators C and Q as follows:

$$Cx = \sum_j \alpha_j (x,e_j)e_j$$

$$Qx = \sum_j (x,e_j)f_{j-1} + Ty.$$

Evidently $T = C + Q$; moreover it is easy to check that C is compact. For consider the finite dimensional operators $C_n = \sum_{j=1}^{n} \alpha_j (x,e_j)e_j$. Then

$$||(C - C_n)x||^2 = \sum_{j>n} |\alpha_j (x,e_j)|^2$$

$$\leq \sup_{j>n} |\alpha_j|^2 \sum_{j>n} |(x,e_j)|^2$$

$$\leq \sup_{j>n} |\alpha_j|^2 \, ||x||^2.$$

Since $\alpha_j \to 0$, we see that C is the uniform limit of the sequence $\{C_n\}$.

It remains to show that Q is quasinilpotent and this requires more subtle arguments. To begin with, since $Q = T - C$, we know that Q is a Reisz operator. Suppose λ_o is a non zero point in $sp(Q)$. Then there exists a non zero z in $N(\lambda_o I - Q)$. Let z be written as

$$z = \sum_j (z,e_j)e_j + w$$

with $w \perp e_j$ for all j. We will show that $w \neq 0$. If w were, in fact, zero, then z would belong to L, the closed subspace spanned by the vectors e_j. Now the definition of Q shows that $Qe_j = f_{j-1} \varepsilon M_{j-1}$ so that $Q^2 e_j \varepsilon M_{j-2}$ etc. and $Q^j e_j = 0$. Now let P_o be the spectral projection corresponding to that part of $sp(Q)$ which lies inside a circle with centre, the origin and radius r less than $|\lambda_o|$. Then a characterization of $R(P_o)$, due to Riesz and Sz.-Nagy [56, p. 424], gives $x \varepsilon R(P_o)$ if and only if

$$\lim\sup ||Q^n x||^{1/n} < r.$$

Hence each e_j belongs to $R(P_o)$ and therefore $L \subseteq R(P_o)$. Now let P denote the spectral projection corresponding to the single point $\{\lambda_o\}$ in $sp(Q)$. Clearly $PP_o = P_oP = 0$ so that the subspace $R(P_o)$ lies inside $N(P)$. Hence $L \subseteq N(P) = R[(\lambda_o I - Q)^p]$ where p is the ascent of $\lambda_o I - Q$. Hence $z \varepsilon N(\lambda_o I - Q) \cap R[(\lambda_o I - Q)^p] = \{0\}$. This shows that $w \neq 0$.

Now $Qz = \lambda_o z$ can be written

$$(\lambda_o I - T)w = \sum_j (z,e_j)f_{j-1} - \lambda_o \sum_j (z,e_j)e_j \varepsilon L.$$

Now $w \perp L$; write L_1 to denote the closed subspace spanned by L and $\{w\}$. Hence $TL_1 \subseteq L_1$ and $(\lambda_o I - T)L_1 \subseteq L$. Let T_1 denote the restriction of T to L_1. Evidently $R[(\lambda_o I - T_1)^k] \subseteq L$ for all k.

Now by Lemma 3.5.1, we know that T_1 is a Riesz operator and $\lambda_o \varepsilon sp(T_1) \subseteq sp(T)$. Also since $L_1 = N[(\lambda_o I - T_1)^q] \oplus R[(\lambda_o I - T_1)^q]$, where q is the ascent of $\lambda_o I - T_1$, we see that

$$N[(\lambda_o I - T_1)^q] \cap (L_1 \backslash L) \neq \emptyset.$$

But obviously

$$N[(\lambda_o I - T_1)^q] \subseteq N[(\lambda_o I - T)^q]$$

so that

$$N[(\lambda_o I - T)^q] \cap (L_1 \backslash L) \neq \emptyset .$$

But $N[(\lambda_o I - T)^q]$ is a subspace of some M_j and hence a subspace of L. This gives the required contradiction.■

REMARK

Further examination of the above proof shows that C is a normal operator and $sp(C) = sp(T)$ with the non zero spectral points having the same spectral multiplicities.

3.6 Extensions and generalizations

Most of the results in this chapter have been extended to the case where T is a closed linear operator. In some cases, the same proofs can be modified, taking into account the additional complications due to the various domains of the operators involved. However, no analogue of the Calkin algebra is available. The following is an attempt to summarize the existing state of affairs.

(3.6.1) FREDHOLM OPERATORS

Substantial sections of the well known monographs by Goldberg [29] and Kato [40] are devoted to Fredholm operators and their generalizations. Often the most useful results involve two operators, the object of the study being to deduce properties of $T + \lambda B$ given certain properties of T and restrictions on the modulus of λ. Such results have natural applications to the perturbation theory of differential operators. The interested reader is referred to the above-named sources for details.

With particular reference to the results of this chapter, it is worth mentioning specific generalizations.

(i) the proof of Lemma 3.2.4 extends without difficulty to the case of closed operators. Any closed operator with $\beta(T)$ finite has closed range.

(ii) the index theorem (3.2.7) can be extended if some modest additional conditions are imposed. In fact, we can have Banach spaces X,Y,Z with $T : X \to Y$ and $S : Y \to Z$ being closed Fredholm operators. If S is densely defined, then, as before,

$$i(ST) = i(S) + i(T)$$

(see [29, Theorem IV.2.7] or [61, Theorem 2.5]).

(iii) Successful extensions of many results to the class of <u>semi-Fredholm operators</u> have been obtained. These are operators which have closed range but only one of $\alpha(T)$ and $\beta(T)$ is required to be finite. In this case, the index may take values $\pm\infty$. Such operators will reappear in a later chapter of this book.

(iv) We showed in Corollary 3.2.9 that, for an infinite dimensional space, operators in $B(X)$ could not have Fredholm region equal to the entire complex plane. It is natural to inquire about the situation for closed operators. Kaashoek and Lay [36] proved that if T is a closed operator with nonempty resolvent set, then $\Phi(T) = \mathbb{C}$ if and only if $(\lambda I - T)^{-1}$ is a Riesz operator for some λ in $\operatorname{res}(T)$. This fact is easily obtained by writing down the operator identity

$$(\mu I - T)(\lambda I - T)^{-1} = (\mu - \lambda)[(\mu - \lambda)^{-1} + (\lambda I - T)^{-1}]$$

and seeing that Theorem 3.2.12 implies the result.

The study of operators with Riesz resolvent is important for applications. Kaniel and Schechter [38] obtained a sufficient condition: Suppose T is a closed operator with dense domain D. Let D^* denote the domain of T^*. Write D_G and D^*_G, respectively, for D and D^* given their graph norms. Then T has the property $\Phi(T) = \mathbb{C}$ if the identity maps $D \to D_G$ and $D^* \to D^*_G$ are both compact. This condition is satisfied for very general elliptic boundary value problems [38, Section 3].

(v) In the study of closed Fredholm operators, most authors have invoked the following lemma which first appeared in [43] (also see Chapter 4 for related material):

If M and N are subspaces of a normed linear space X with $\dim M > \dim N$ (so that $\dim N < \infty$), then there exists $m \neq 0$ in M such that $||m|| = \inf_{n \in N} ||m - n||$.

This lemma, easy to prove when X is a Hilbert space, apparently needs Borsuk's antipodal mapping theorem for its proof in the general case. However, it is interesting to study M. Schechter's development of the subject in [61] where he avoids the use of the above lemma and yet is still able to obtain essentially all the known results.

(vi) In addition to the local result, Theorem 3.2.10, it is possible to show that the index of an operator is constant on each component of the Fredholm set. The proof of this is quite analogous to that of Theorem 3.2.11; suppose λ_1 and λ_2 belong to the same component Ω of $\Phi(T)$ with $i(\lambda_1 I - T) \neq i(\lambda_2 I - T)$. Join λ_1 to λ_2 with curve Γ lying in Ω. Then use Theorem 3.2.10 and the compactness of Γ to deduce the result.

(3.6.2) ASCENT AND DESCENT

A comprehensive study of ascent and descent is given by Taylor [65] with a strong emphasis on purely algebraic methods. This work was later augmented by Kaashoek [35] who settled certain questions raised by Taylor, and also by Lay [44] who put many parts of this area in a rather definitive form. It should be noted that these writers follow Taylor in writing $\alpha(T)$ and $\delta(T)$ respectively, to denote ascent and descent while using $n(T)$ and $d(T)$ for the dimension of the nullspace and the codimension of the range.

In particular, Lay showed that Lemma 3.4.2 of this chapter is valid for a closed operator if either the resolvent set is non empty or the Fredholm region is non empty and the operator is densely defined. Substantial parts of our proofs are modelled on his. He also obtained a variety of interesting conditions sufficient to ensure that a given complex number is a pole of the resolvent. In addition, he derives the characterization of Riesz operators:

An operator T in $B(X)$ is a Riesz operator if and only if $\lambda I - T$ is semi Fredholm for every non zero λ.

This result is also implicit in an earlier work of Kato [39].

As an analogue of the Fredholm region, it is profitable to study the <u>Riesz region</u> of an operator, defined as the set of complex λ such that $\lambda I - T$ has finite ascent and descent. Suppose we write $\mathcal{R}(T)$ for the Riesz region of operator T. Then it follows from the above mentioned generalization of Lemma 3.4.2 that $\mathcal{R}(T) \cap sp(T)$ consists of poles of the resolvent operator and that $\mathcal{R}(T)$ is an open set.

Another contribution worthy of serious attention is that of P. Saphar [58] whose definition of ascent and descent makes allowance for different ordinal values. He then develops the theory of operators T in $B(X)$ which are

(i) "regular" i.e. for some $S \varepsilon B(X)$, $TST = T$ and $STS = S$. Such an S is called a relative inverse of T. (It is easy to show that T has a relative inverse if and only if $R(T)$ and $N(T)$ are closed, complemented subspaces).

(ii) "perfect" i.e. if δ is the ordinal-valued descent of T, then $T^{-1}R(T^{\delta}) = R(T^{\delta})$ where the symbol "T^{-1}" is merely to be understood in the sense of preimage. It is not difficult to show that this condition is equivalent to $R(T^{\delta}) \geq N(T)$.

Credit is also due to H. Heuser whose inaugural dissertation [33] of 1956 predates most other work in the field and contains many results rediscovered later. In addition, he proves the following useful facts for T in $B(X)$.

(a) if at least one of the quantities $\alpha(T)$, $\beta(T)$ is finite, then $a(T) < \infty$ implies $\beta(T) \geq \alpha(T)$ and $d(T) < \infty$ implies $\beta(T) \leq \alpha(T)$

(b) if $\alpha(T) = \beta(T) < \infty$, then $a(T)$ is finite if and only if $d(T)$ is finite.

(3.6.3) ALGEBRAIC PROPERTIES OF RIESZ OPERATORS

Since the sum and product of commuting quasinilpotent elements of a Banach algebra are again quasinilpotent, it is easy to see from Theorem 3.3.1 that if T and S belong to $R(X)$ and T commutes with S then $T + S$ and TS are both Riesz operators. It is not difficult to find examples showing that this conclusion is false in general without the commutativity assumption. We may take the space X^2 and the operator T defined in 3.1 of this chapter and also define $S \varepsilon B(X^2)$ by writing $S(x,y) = (0,x)$. Then both S and T are Riesz operators but none of $S + T$, ST and TS are of this type.

We might also mention in the case of commuting operators T and S, if one of the pair belongs to $R(X)$, then the product is a Riesz operator.

Finally, if T is a Riesz operator and f is analytic on an open set containing $sp(T)$ with $f(0) = 0$, then $f(T)$ is a Riesz operator. This fact is a straight forward consequence of Theorem 3.3.1 and the relation $\pi f(T) = f(\pi(T))$ where π is the canonical mapping of $B(X)$ into the Calkin algebra $C(X)$.

(3.6.4) THE WEST DECOMPOSITION

West [66] calls a Riesz operator <u>fully decomposable</u> if it can be written as $T = C + Q$ with C compact, Q quasinilpotent and $CQ = QC = 0$. He shows that if the spectral projections P_j corresponding to the non zero points of $sp(T)$ can be enumerated in such a way that $\sum_1^\infty TP_j$ is convergent in $B(X)$, then T is fully decomposable.

However, Gillespie and West [24] showed that, even in Hilbert space, some Riesz operators are not fully decomposable; in fact, they produced an example in which no decomposition $T = C + Q$ exists with C and Q commuting. We will give the details of this interesting example.

Let $H = \ell_2$ with the standard orthonormal basis $\{e_j\}_1^\infty$. Define T on the basis elements as follows

$$Te_{2j} = \frac{1}{2j} e_{2j}$$

$$Te_{2j+1} = \frac{1}{2j+1} e_{2j+1} + e_{2j+2}.$$

Then it is easy to verify that T^2 is compact so that T is Riesz (since $\pi(T^2) = \pi(T)^2 = 0$). However T is not compact since the sequence $\{Te_{2j+1}\}_{j=1}^\infty$ does not contain a convergent subsequence. Now the spectrum of T consists of zero and the points $\lambda_j = \frac{1}{j}$ with the corresponding projections having ranges E_j which are one dimensional

$$E_{2j+1} = \text{span}\{e_{2j+1} + (2j+1)(2j+2)e_{2j+2}\}$$

$$E_{2j} = \text{span}\{e_{2j}\}.$$

Now suppose T could be decomposed: $T = C + Q$ with $C \in K(H)$, Q quasinilpotent and $CQ = QC$. Then T would commute with Q. If $x \in E_j$, then it is easy to verify that $Tx = \frac{1}{j}x$ so

$$TQx = QTx = Q\,\frac{1}{j}x = \frac{1}{j}\,Qx.$$

Hence E_j is a one-dimensional invariant subspace for T so $sp(Q|E_j)$ must consist of a single eigenvalue. But by Lemma 3.5.1 (ii), $sp(Q|E_j) \subseteq sp(Q) = \{0\}$ so that $Q|E_j$ is a one-dimensional operator with spectrum $\{0\}$. Hence $Q|E_j = 0$. But the subspaces E_j span H. Thus Q must be the zero operator. But this contradicts the fact that $T \notin K(H)$.

It is perhaps of interest to note that if we take the obvious decomposition of T into $C + Q$ given by $Ce_j = \frac{1}{j} e_j$ and $Qe_{2j} = 0$, $Qe_{2j+1} = e_{2j+2}$ then C is compact, Q quasinilpotent and of course $QC \neq CQ$. But $QC - CQ$ is quasinilpotent; in fact, by direct calculation

$$(QC - CQ)e_{2j} = 0$$

$$(QC - CQ)e_{2j+1} = \frac{1}{(2j + 1)(2j + 2)} e_{2j+2}$$

so that $(QC - CQ)^2 = 0$.

It should be worthwhile constructing other examples to see whether one can get evidence for the obvious conjecture that a decomposition $T = C + Q$ can always be obtained with $CQ - QC$ a quasinilpotent operator.

Chapter 4

SEMI-FREDHOLM OPERATORS

4.1 Introduction

With Theorem 3.2.8 we proved that the Fredholm operators, $\Phi(X)$, in $B(X)$ were equal to

$$\pi^{-1}(G)$$

the inverse image of the invertible elements in $B(X)/K(X)$.

In this chapter we wish to look at the classes of semi-Fredholm operators (Definition 1.3.1) in $B(X)$ and attempt to classify them in terms of the mapping π. We will also look at Fredholm and semi-Fredholm operators under perturbations of various kinds. Some of these results will be used in the development of Chapter 5.

4.2 Semi-Fredholm operators as open semi groups in $B(X)$

We now present results which will prove that the semi groups $\Phi_+(X)$ and $\Phi_-(X)$ (see Corollary 1.3.3) are open in $B(X)$. The reader is referred to I. Gohberg and M. Krein [27] for an extended discussion and reference to sources.

(4.2.1) THEOREM

If $T \in \Phi_+(X)$ then there exists an $\varepsilon > 0$ such that $U \in B(X)$ and $||U|| < \varepsilon$ implies $T + U \in \Phi_+(X)$. Moreover ε can be chosen so that

$$\alpha(T + U) \leq \alpha(T) \text{ and if } \beta(T) = \infty \text{ then } \beta(T + U) = \infty.$$

Proof. $N(T)$ is finite dimensional, so there exists a closed subspace Q such that $X = N(T) \oplus Q$. Then T_Q (recall T_Q denotes the restriction of T to Q) is one to one and has closed range $R(T)$, so T_Q^{-1} is continuous and therefore bounded. Hence there exists $\delta > 0$ such that $||T(x)|| \geq \delta||x||$ for all $x \in Q$.

Let $\varepsilon = \frac{\delta}{3}$. If $U \in B(X)$ with $||U|| < \varepsilon$, we have

$$0 \le ||U(x)|| < \varepsilon\ ||x|| = \frac{\delta}{3}\ ||x|| \quad \text{for all} \quad x \in Q.$$

From this we get

$$(1) \quad ||(T + U)x|| \ge ||T(x)|| - ||U(x)|| \ge \delta||x|| - \frac{\delta}{3}\ ||x|| = \frac{2\delta}{3}\ ||x||$$

$$\text{for all} \quad x \in Q.$$

So $(T + U)_Q^{-1}$ exists and is continuous. Now $T + U$ is one to one on Q and $R[(T + U)_Q]$ is closed. We must show that $\alpha(T + U) < \infty$ and $R(T + U)$ is closed. Let $\alpha(T) = p$. Suppose $N(T + U)$ had $p + 1$ linearly independent elements, $x_1, x_2, \ldots, x_{p+1}$. Then since $N(T + U) \cap Q = \{0\}$ we would have that $x_1, x_2, \ldots, x_{p+1}$ are linearly independent modulo Q. Since $X = N(T) \oplus Q$, it follows that X/Q is p dimensional, so there cannot exist $p+1$ elements of X which are linearly independent modulo Q. This contradiction implies that $\alpha(T + U) \le \alpha(T)$.

Now to show that $R(T + U)$ is closed. Since $X = [Q \oplus N(T + U)] + N(T)$, there exists a finite dimensional subspace K such that

$$X = [Q \oplus N(T + U)] \oplus K.$$

Therefore

$$R(T + U) = (T + U)K + (T + U)(Q \oplus N(T + U)).$$

Since $(T + U)K$ is finite dimensional, it remains to show that $(T + U)(Q \oplus N(T + U))$ is closed. But this reduces to showing that $(T + U)Q$ is closed and we noted this earlier.

Now to get that $\beta(T) = \beta(T + U)$ if $\beta(T) = \infty$.

$$||(T + U)x - T(x)|| = ||U(x)|| \le \varepsilon\ ||x|| = \frac{\delta}{3}\ ||x|| \le \frac{1}{3}\ ||T(x)||$$

$$\text{for all} \quad x \in Q,$$

and using (1) we get

$$||T(x) - (T + U)x|| = ||U(x)|| \le ||U||\ ||x|| \le \frac{3||U||}{2\delta}\ ||(T + U)x||$$

$$\text{for all} \quad x \in Q.$$

Note that $\dfrac{3||U||}{2\delta} < \dfrac{3\varepsilon}{2\delta} = \dfrac{1}{2}$. So the above two inequalities give us an estimate of the gap between $R_1 = R[(T + U)_Q]$ and $R_2 = R[T_Q]$ of $\Theta(R_1,R_2) < \frac{1}{2}$.

By Appendix 1 we get that

$$\dim R_1^{\perp} = \dim R_2^{\perp} = \beta(T), \quad \text{since} \quad R_2 = R(T).$$

Inasmuch as $R(T + U) = R_1 \oplus W$, where W is a suitable finite dimensional subspace, we see that $\beta(T + U) = \infty$ if $\beta(T) = \infty$.■

(4.2.2) THEOREM

If $T \in \Phi_-(X)$ then there exists an $\varepsilon > 0$ such that $U \in B(X)$ and $||U|| < \varepsilon$ implies $T + U \in \Phi_-(X)$. Moreover ε can be chosen so that $\beta(T + U) \leq \beta(T)$ and if $\alpha(T) = \infty$ then $\alpha(T + U) = \infty$.

Proof. Use the relationships between T and T^* of section 1.3.■

4.3 Semi-Fredholm operators and the mapping $\pi: B(X) \to B(X)/K(X)$

In this section we show that there is a connection between certain of the semi-Fredholm operators and the classes of right and left invertible elements in the quotient algebra $B(X)/K(X)$. This material first appeared in B. Yood [68]. We being with a useful lemma.

(4.3.1) LEMMA

If $T \in \Phi_+(X)$ and E is any closed subspace of X, then $T(E)$ is closed.

Proof. Suppose E is a closed subspace of X. Let σ be the natural map, $\sigma: X \to X/N(T)$. By the open mapping theorem [22], σ is an open mapping. If we define $\hat{T}: (X/N(T)) \to X$ by $\hat{T}(\sigma(x)) = T(x)$, then it is clear that $\hat{T}$ is one to one, bounded and onto $R(T)$, so $\hat{T}^{-1}$ is bounded. Therefore since $T(E) = \hat{T}\sigma(E)$ with σ open and $\hat{T}$ and isomorphism, it follows that $T(E)$ is closed.■

(4.3.2) THEOREM

Let $T \in B(X)$. There exists $U \in B(X)$ and $K \in K(X)$ such that $TU = I + K$ if and only if $T \in \Phi_-(X)$ and there exists a bounded projection of X onto $N(T)$.

Proof. (→) Suppose $TU = I + K$ for some $U \in B(X)$ and $K \in K(X)$. $I + K \in \Phi(X)$ so by Corollary 1.3.6 we have $U \in \Phi_+(X)$ and $T \in \Phi_-(X)$. We have $X = N(I + K) \oplus W$ for some closed subspace W and $N(U) \subset N(I+K)$, so U is one to one on W. By Lemma 4.3.1 we see that $U(W)$ is closed.

Let $x \in N(T) \cap U(W)$. Then $T(x) = 0$ and $U(y) = x$ for some $y \in W$. Therefore $TU(y) = 0$ or $(I + K)y = 0$ and since $y \in W$ we conclude that $y = 0$ and therefore $x = 0$. Let us examine $N(T) \oplus U(W)$. Since $T(N(T) \oplus U(W)) = TU(W) = (I + K)(W)$ is closed, we see that $T^{-1}[(I + K)W] = \{x | T(x) \in (I + K)W\} = N(T) \oplus U(W)$ is closed. We claim that $X/[N(T) \oplus U(W)]$ is finite dimensional. Let $x_1, x_2, \ldots, x_s \in X$ be linearly independent modulo $N(T) \oplus U(W)$. Then $Tx_1, \ldots, Tx_s$ are linearly independent modulo $TU(W) = (I + K)W = (I + K)X$. But $\beta(I + K) < \infty$, so $s \leq \beta(I + K) < \infty$. Therefore there exists a finite dimensional subspace F such that $X = N(T) \oplus (U(W) + F)$ is closed, we conclude that there exists a bounded projection of X onto $N(T)$ (Lemma 2.5.1).

(←) Suppose $T \in \Phi_-(X)$ and there exists a continuous projection of X onto $N(T)$. Write $X = N(T) \oplus W$, where W is a closed subspace. Since $T(W) = R(T)$, $T(W)$ is closed and so T_W^{-1} is a continuous map of $R(T)$ onto W. $T \in \Phi_-(X)$ implies $R(T) \oplus F = X$ for a finite dimensional space F. Therefore by Lemma 2.5.1 there exists a bounded projection P of X onto $R(T)$. Now $T_W^{-1}P \in B(X)$ so consider $T(T_W^{-1}P)$. $N(TT_W^{-1}P) = N(P) = F$ is finite dimensional and $R(TT_W^{-1}P) = R(T)$ has finite deficiency with $R(T) \oplus F = X$. So $TT_W^{-1}P \in \Phi(X)$ and by Lemma 3.2.6 there exists an $S \in B(X)$ such that

$$T(T_W^{-1}PS) = I + K \quad \text{for some} \quad K \in K(X). \blacksquare$$

(4.3.3) THEOREM

Let $T \in B(X)$. There exists $V \in B(X)$ and $K \in K(X)$ such that $VT = I + K$ if and only if $T \in \Phi_+(X)$ and there exists a bounded projection of X onto $R(T)$.

Proof. (→) Suppose $VT = I + K$ for some $V \in B(X)$ and $K \in K(X)$. $I + K \in \Phi(X)$ so by Corollary 1.3.6 we have $T \in \Phi_+(X)$ and $V \in \Phi_-(X)$. As in the proof of Theorem 4.3.2 we have $X = N(I + K) \oplus W$ for a closed subspace W. Also $T(W)$ is closed, $N(V) \oplus T(W)$ is closed and $X/[N(V) \oplus T(W)]$ is finite dimensional. Since $X = N(I + K) \oplus W$, we get

(1) $$R(T) = T[N(I + K)] + T(W).$$

Then $N(V) + R(T) = [N(V) \oplus T(W)] + T[N(I + K)]$ with $N(V) \oplus T(W)$ closed and $T[N(I + K)]$ finite dimensional. Therefore $N(V) + R(T)$ is closed.

Since $N(V) \cap T(N(I + K))$ is finite dimensional and $N(V)$ is closed there exists a closed subspace Z such that

(2) $$N(V) = [N(V) \cap T(N(I + K))] \oplus Z.$$

From (1) and (2) we see that

(3) $$N(V) + R(T) = R(T) + Z.$$

We show that the sum in (3) is direct. For suppose that $w \in R(T) \cap Z$ where $w = T(x)$. Inasmuch as $Z \subset N(V)$, we get $0 = V(w) = VT(x) = (I + K)(x)$ Then we see that w lies in $N(V) \cap T(N(I + K))$ as well as in Z. It follows now from (2) that $w = 0$.

Inasmuch as $\beta[N(V) + R(T)] < \infty$, there exists a finite dimensional subspace F such that

$$X = R(T) \oplus Z \oplus F.$$

But $Z \oplus F$ is closed. Therefore, by Lemma 2.5.1, there exists a bounded projection of X onto $R(T)$.

($\longleftarrow$) Suppose $T \in \Phi_+(X)$ and there exists a bounded projection P of X onto $R(T)$. $N(T)$ is finite dimensional so $X = N(T) \oplus E$ for some closed subspace E. Now T_E is one to one and $T(E) = R(T)$ is closed, so T_E^{-1} is continuous from $R(T)$ onto E. It is straightforward to show that $T_E^{-1}PT \in \Phi(X)$ and therefore, by Lemma 3.2.6, there exists $S \in B(X)$ and $K \in K(X)$ such that $(ST_E^{-1}P)T = I + K$.■

(4.3.4) SUMMARY

Let us define the following subclasses of $\Phi_+(X)$ and $\Phi_-(X)$:

$\Phi_r(X) = \{T \in \Phi_-(X) |$ there exists a bounded projection of X onto $N(T)\}$

$\Phi_\ell(X) = \{T \in \Phi_+(X) |$ there exists a bounded projection of X onto $R(T)\}$.

Let G_r and G_ℓ be the right and left, respectively, invertible elements of $B(X)/K(X)$. By Theorems 4.3.2 and 4.3.3 we conclude that

$$\Phi_r(X) = \pi^{-1}(G_r)$$

and

$$\Phi_\ell(X) = \pi^{-1}(G_\ell)$$

where π is the natural map, $\pi : B(X) \to B(X)/K(X)$.

4.4 Perturbations and restrictions of semi-Fredholm and Fredholm operators

We begin with two classic theorems about Fredholm operators. For original sources see [1] and [68].

(4.4.1) THEOREM

If $T \in \Phi(X)$ then there exists $\varepsilon > 0$ such that for each $U \in B(X)$ with $||U|| < \varepsilon$ we have $T + U \in \Phi(X)$ and $i(T + U) = i(T)$. Therefore the index is continuous on the open semigroup $\Phi(X)$.

Proof. Suppose $T \in \Phi(X)$. By Theorems 4.2.1 and 4.2.2 there exists a $\delta > 0$ such that for each $U \in B(X)$ with $||U|| < \delta$ we have $T + U \in \Phi_-(X) \cap \Phi_+(X) = \Phi(X)$. By Lemma 3.2.6 and its corollary there exist $T_o \in \Phi(X)$ and $K \in K(X)$ such that $TT_o = I + K$. By Theorem 3.2.7 we have $i(T) + i(T_o) = i(I + K) = 0$. Now $(T + U)T_o = TT_o + UT_o = I + K + UT_o$. By Theorem 2.1.3, $(I + UT_o)^{-1}$ exists if we choose $||U||$ small enough. Choose $\varepsilon > 0$ such that $0 < \varepsilon < \delta$ and where $(I + UT_o)^{-1}$ exists. Now for such U,

$$(I + UT_o)^{-1}(T + U)T_o = I + (I + UT_o)^{-1}K, \quad \text{if} \quad ||U|| < \varepsilon$$

where $(I + UT_o)^{-1} K \in K(X)$ so by Theorem 1.4.7 we have $i(I + (I + UT_o)^{-1}K) = 0$ and by Theorem 3.2.7 $i[(I + UT_o)^{-1}] + i(T + U) + i(T_o) = 0$. But $i[(I + UT_o)^{-1}] = 0$ since its null space is $\{0\}$ and its range is X. We have shown that $i(T) + i(T_o) = 0$ and $i(T + U) + i(T_o) = 0$, so we conclude that $i(T) = i(T + U)$ for all $U \in B(X)$ with $||U|| < \varepsilon$.■

(4.4.2) THEOREM

If $T \in \Phi(X)$ and $K \in K(X)$ then $T + K \in \Phi(X)$ and $i(T + K) = i(T)$.

Proof. Suppose $T \in \Phi(X)$. By Lemma 3.2.6 there exist bounded operators T_1 and T_2 and compact operators K_1 and K_2 such that

$$T_1T = I + K_1$$

and

$$TT_2 = I + K_2.$$

Therefore for any $K \varepsilon K(X)$ we have

$$T_1(T + K) = T_1T + T_1K = I + (K_1 + T_1K) = I + K_3, \quad K_3 \varepsilon K(X)$$

and

$$(T + K)T_2 = TT_2 + KT_2 = I + (K_2 + KT_2) = I + K_4, \quad K_4 \varepsilon K(X)$$

so applying Lemma 3.2.6 we conclude that $T + K \varepsilon \Phi(X)$. Now using the same method as we did in the proof of Theorem 4.4.1 we may conclude that

$$i(T_1T) = 0, \quad \text{so} \quad i(T_1) + i(T) = 0$$

and

$$i[T_1(T + K)] = 0, \quad \text{so} \quad i(T_1) + i(T + K) = 0.$$

But these equations imply that $i(T) = i(T + K)$.■

We conclude this section by developing some specialized results which we will subsequently need in Chapter 5.

(4.4.3) DEFINITION

An operator $T \varepsilon B(X)$ is said to be <u>strictly singular</u> if, for every infinite dimensional closed subspace $M \subset X$, the restriction of T to M is not a homeomorphism. Let $S(X)$ denote the set of all strictly singular operators on X.

For a detailed study of the properties of strictly singular operators, we refer the reader to [29]. We remark, however, that $S(X)$ is a closed ideal. We are interested in studying the relationships between strictly singular and semi-Fredholm operators and between $S(X)$ and other ideals in $B(X)$.

(4.4.4) THEOREM

$$K(X) \subset S(X) \text{ for any Banach space } X.$$

Proof. Suppose $T \varepsilon B(X)$ and $T \notin S(X)$. Then there exists an infinite dimensional closed subspace $M \subset X$ such that T_M is a homeomorphism. Let $\{x_n\}$ be a sequence in the unit ball of M containing no Cauchy subsequence. Then $\{T(x_n)\}$ contains no Cauchy subsequence since T_M is a homeomorphism and therefore $T \notin K(X)$. This conclusion implies that $K(X) \subset S(X)$.■

(4.4.5) LEMMA

Let $T \in B(X)$ and suppose $R(T)$ is not closed in X. Then for each $\varepsilon > 0$ there exists an infinite dimensional closed subspace Z of X such that $||T_Z|| < \varepsilon$.

Proof. First we prove that if Y is any closed subspace of X with finite deficiency, then T does not have a bounded inverse on Y. If we assume T has a bounded inverse on Y, then surely T(Y) is closed. Now $X = Y \oplus N$ where N is a finite dimensional subspace of X. Hence $T(X) = T(Y) + T(N)$ with T(Y) closed and T(N) finite dimensional, so T(X) is closed, a contradiction.

Since T does not have a bounded inverse on X, there exists an $x_1 \in X$ such that $||x_1|| = 1$ and $||Tx_1|| < \frac{\varepsilon}{3}$. By the Hahn-Banach theorem there exists an $x_1^* \in X^*$ such that $||x_1^*|| = 1$ and $x_1^*(x_1) = ||x_1|| = 1$. Since $N(x_1^*)$ has deficiency 1 in X, there exists an $x_2 \in N(x_1^*)$ such that $||x_2|| = 1$ and $||Tx_2|| < \frac{\varepsilon}{3^2}$. There exists an $x_2^* \in X^*$ such that $||x_2^*|| = 1$ and $x_2^*(x_2) = ||x_2|| = 1$. Since $N(x_2^*) \cap N(x_1^*)$ has finite deficiency in X, there exists an $x_3 \in N(x_1^*) \cap N(x_2^*)$ such that $||x_3|| = 1$ and $||Tx_3|| < \frac{\varepsilon}{3^3}$. Inductively, sequences $\{x_k\} \subset X$ and $\{x_k^*\} \subset X^*$ are constructed so that

(1) $||x_k|| = ||x_k^*|| = x_k^*(x_k) = 1$ and $||Tx_k|| < \frac{\varepsilon}{3^k}$ for $k = 1,2,\ldots$

(2) $x_i^*(x_k) = 0$ for $i < k$.

It is easily verified that the set $\{x_k\}$ is linearly independent. Hence M = the span of $\{x_1,x_2,\ldots\}$ is an infinite dimensional subspace of X. We will show that T_M (and therefore $T_{\bar{M}}$) has norm not exceeding ε.

Suppose $x = \sum_{i=1}^{m} a_i x_i \in M$. Then from (1) and (2)

$$|a_1| = |x_1^*(x)| \leq ||x_1^*||\ ||x|| = ||x||.$$

In fact

(3) $|a_k| \leq 2^{k-1}||x||$ for $k = 1,\ldots,m.$

For suppose (3) is true for $k \leq j < m$. Then, from (1) and (2), we get

$$x^*_{j+1}(x) = \sum_{i=1}^{j} a_i x^*_{j+1}(x_i) + a_{j+1}.$$

Hence using the induction hypothesis, we have

$$|a_{j+1}| \leq |x^*_{j+1}(x)| + \sum_{i=1}^{j} |a_i|\ |x^*_{j+1}(x_i)| \leq ||x|| + \sum_{i=1}^{j} 2^{i-1}||x|| \leq 2^j||x||.$$

Thus (3) follows by induction and we have

$$||T(x)|| \leq \sum_{i=1}^{m} |a_i|\ ||T(x_i)|| \leq \sum_{j=1}^{m} 2^{i-1}\frac{\varepsilon}{3^i}\ ||x|| = \frac{\varepsilon}{2}\ ||x||.$$

Since $x \in M$ was arbitrary we conclude that $||T_M|| \leq \frac{\varepsilon}{2}$ and therefore $||T_{\bar{M}}|| \leq \frac{\varepsilon}{2} < \varepsilon$.■

(4.4.6) THEOREM

If $T \in \Phi_+(X)$ and $U \in S(X)$, then $T + U \in \Phi_+(X)$.

Proof. We know $X = N(T) \oplus W$ and, as in our earlier arguments, T_W^{-1} is continuous from $R(T)$ onto W. Now $R(T + U) = (T + U)(N(T)) + (T + U)(W)$ and, since $(T + U)(N(T))$ is finite dimensional, we will know $R(T + U)$ is closed if $(T + U)(W)$ is closed. Now since T_W^{-1} is continuous there exists a $k > 0$ such that $||T(x)|| \geq k||x||$ for all $x \in W$. If $(T + U)(W)$ is not closed we apply Lemma 4.4.5 to get an infinite dimensional closed subspace $Z \subset W$ such that $||(T + U)_Z|| < \frac{k}{2}$. We would then have for each $x \in Z$ that

$$||U(x)|| \geq ||T(x)|| - ||(T + U)x|| \geq k||x|| - \frac{k}{2}\ ||x|| = \frac{k}{2}\ ||x||,$$

which implies that U is a homeomorphism on Z. But this is a contradiction to $U \in S(X)$. Therefore $(T + U)(W)$ is closed and so $R(T + U)$ is closed.

Lastly we must show that $\alpha(T + U) < \infty$. $N(T + U) \cap N(T)$ is finite dimensional so there exists a closed subspace Y such that $N(T + U) = [N(T + U) \cap N(T)] \oplus Y$. By Lemma 4.3.1 we know $T(Y)$ is closed. We also know T is one to one on W, so T_Y^{-1} is continuous. Now $U = -T$ on $N(T + U)$ so U must also be a homeomorphism on Y. But $U \in S(X)$ so Y must be finite dimensional. This implies that $N(T + U)$ is finite dimensional.■

The following results can be found in M. Schechter [62] and M. Schechter and A. Lebow [45].

(4.4.7) THEOREM

For $T \in B(X)$ the following statements are equivalent:

(i) $T \in \Phi_+(X)$.

(ii) $\alpha(T - K) < \infty$ for all $K \in K(X)$.

(iii) For every infinite dimensional closed subspace $M \subset X$, T_M is not a compact operator.

Proof. We prove that (i) is equivalent to (ii). It is then an easy exercise to show that (iii) is equivalent to (i) and (ii). It is interesting to note that the equivalence of (ii) and (iii) appears to be nontrivial without knowing the equivalence of (i) and (ii).

($\rightarrow$) Suppose $T \in \Phi_+(X)$, then since by Theorem 4.4.4, $K(X) \subset S(X)$, it follows by Theorem 4.4.6 that $T - K \in \Phi_+(X)$ for all $K \in K(X)$.

($\leftarrow$) Suppose $T \notin \Phi_+(X)$. Let us suppose for the moment that there are sequences

$$\{x_k\} \subset X, \quad \{x_k^*\} \subset X^* \text{ such that } ||x_k|| = 1, \quad ||x_k^*|| \leq 2^{k-1}, \tag{4.4.8}$$
$$x_j^*(x_k) = \delta_{jk} \text{ and } ||Tx_k|| \leq 2^{1-2k}.$$

Let
$$K_n(x) = \sum_1^n x_k^*(x)T(x_k) \quad \text{for } n = 1,2,\ldots.$$

Then for $n > m$

$$||(K_n - K_m)x|| \leq \sum_{m+1}^{n} 2^{k-1}\, 2^{1-2k}||x||,$$

which shows that $||K_n - K_m|| \to 0$ as $m, n \to \infty$. Therefore $K_n \to K$, where

$$K(x) = \sum_1^\infty x_k^*(x)T(x_k).$$

Now $K(x) = T(x)$ for any $x = x_k$ and also for any linear combination of the x_k. Since the x_k are linearly independent, it follows that $\alpha(T - K) = \infty$. Therefore we are done if we can show sequences as in (4.4.8) exist.

Surely T does not have a bounded inverse. Thus there exists a vector $x_1 \in X$ with $||x_1|| = 1$ such that $||T(x_1)|| \le \frac{1}{2}$. By the Hahn-Banach Theorem there exists an $x_1^* \in X^*$ such that $||x_1^*|| = 1$ and $x_1^*(x_1) = 1$. Suppose we have constructed a biorthogonal system $\{x_k\}$ and $\{x_k^*\}$ for $k = 1,2,\ldots,n-1$ such that $||x_k|| = 1$, $||T(x_k)|| \le 2^{1-2k}$, $||x_k^*|| \le 2^{k-1}$ for each k.

Since the restriction of T to the closed subspace $N = \bigcap_{k=1}^{n-1} N(x_k^*)$ cannot have a bounded inverse, there is a vector $x_n \in N$ such that $||x_n|| = 1$ and $||T(x_n)|| < 2^{1-2n}$. Let $g \in X^*$ be any functional such that $g(x_n) = 1$ and $||g|| = 1$. Then the functional

$$x_n^* = g - \sum_{k=1}^{n-1} g(x_k)x_k^* \quad \text{has the properties}$$

$$x_n^*(x_k) = \delta_{nk} \quad \text{for } k = 1,2,\ldots,n \quad \text{and} \quad ||x_n^*|| \le 2^{n-1}.$$

Therefore by induction we have shown the existence of sequences satisfying equations (4.4.8).■

(4.4.9) NOTATION

To this point for $T \in B(X,Y)$ we have not needed a definition of $\beta(T)$ if $R(T)$ is not closed. In the following we use the general definition that $\beta(T)$ is the dimension of $Y/\overline{R(T)}$.

(4.4.10) THEOREM

$T \in \Phi_-(X)$ if and only if $\beta(T - K) < \infty$ for all $K \in K(X)$.

Proof. ($\rightarrow$) Suppose $T \in \Phi_-(X)$, and let $K \in K(X)$. Then $K^* \in K(X^*)$ by Theorem 1.2.1 and $T^* \in \Phi_+(X^*)$. By Theorems 4.4.4 and 4.4.6 we now have $T^* - K^* \in \Phi_+(X^*)$. Therefore $T - K \in \Phi_-(X)$.

($\leftarrow$) Suppose $T \notin \Phi_-(X)$. Then either $R(T)$ is closed and $\beta(T) = \infty$ or $R(T)$ is not closed. In the first case, there exists a $K \in K(X)$ (namely $K = 0$) such that $\beta(T - K) = \infty$ and we are finished. So assume $R(T)$ is not closed. Let $\{a_n\}$ be the sequence of integers defined inductively by

$$a_1 = 2, \quad a_n = 2(1 + \sum_{1}^{n-1} a_k), \quad n = 2,3,\ldots.$$

We claim that there are two sequences $\{y_k\} \subset X$ and $\{y_k^*\} \subset X^*$ such that

$$||y_k|| \le a_k, \quad ||y_k^*|| = 1, \quad ||T^*(y_k^*)|| < \frac{1}{2^k a_k}, \quad \text{and}$$

$$(4.4.11) \qquad y_j^*(y_k) = \delta_{jk}, \quad j,k = 1,2,\ldots.$$

Assuming these sequences exist we define the finite rank operators

$$K_n(x) = \sum_{1}^{n} T^*y_k^*(x)y_k, \quad n = 1,2,\ldots.$$

Then for $n > m$

$$||K_n(x) - K_m(x)|| \le \sum_{m+1}^{n} ||T^*y_k^*|| \, ||x|| \, ||y_k|| \le (\sum_{m+1}^{n} \frac{1}{2^k})||x|| \le ||x||/2^m.$$

Therefore K_n converges to the compact operator $K(x) = \sum_{1}^{\infty} T^*y_k^*(x)y_k$. Now for each $x \in X$ and each k we have

$$y_k^*(Kx) = T^*y_k^*(x) = y_k^*(Tx).$$

Consequently each of the y_k^* annihilates $R(T - K)$. Since the y_k^* are linearly independent, it follows that $\beta(T - K) = \infty$.

It remains to find sequences as mentioned above. We use induction. Since $R(T)$ is not closed, the same is true of $R(T^*)$. Hence there exists y_1^* such that $||y_1^*|| = 1$ and $||T^*(y_1^*)|| < ¼$ and there exists y_1 such that $||y_1|| < 2$ with $y_1^*(y_1) = 1$. Now assume that $y_1, y_2, \ldots, y_{n-1}$ and $y_1^*, y_2^*, \ldots, y_{n-1}^*$ have been found satisfying equations (4.4.11). Then there exists an annihilator y_n^* of $y_1, y_2, \ldots, y_{n-1}$ such that $||y_n^*|| = 1$ and $||T^*(y_n^*)|| < \frac{1}{2^n a_n}$. There also exists $y \in X$ such that $y_n^*(y) = 1$, $||y|| < 2$. Let

$$y_n = y - \sum_1^{n-1} y_k^*(y) y_k.$$

Then

$$||y_n|| \leq ||y||(1 + \sum_1^{n-1} ||y_k||) \leq 2(1 + \sum_1^{n-1} a_k) = a_n$$

by the induction hypothesis. Moreover $y_n^*(y_n) = 1$, $y_n^*(y_k) = 0$ for $1 \leq k < n$ by the way y_n^* and y_n were chosen. We also have

$$y_k^*(y_n) = y_k^*(y) - y_k^*(y) = 0 \quad \text{for } 1 \leq k < n.$$

Therefore equation (4.4.11) holds for all n, and the proof is complete.■

(4.4.12) REMARKS

Many of the results of this chapter can be done for $B(X,Y)$ with little change required in the proofs. More detailed results are also available for $\Phi_-(X,Y)$, $\Phi_+(X,Y)$ and $\Phi(X,Y)$, see [45] and [62].

Chapter 5

IDEAL THEORY FOR B(X)

5.1 Introduction

In this chapter we investigate the ideal structure of B(X). We will be principally concerned with norm closed ideals, but we also give some attention to the ideal of finite dimensional operators, F(X). It is to be understood throughout that ideal means two sided ideal.

It is natural to ask how complicated the ideal structure can be and we shall demonstrate a Banach space X where B(X) is extremely complicated. To the other extreme, we shall prove that for $X = \ell_p$ $(1 \leq p < \infty)$ or c_o the ideal of compact operators is the unique nonzero proper closed ideal of B(X).

Finally we look at the question of containment of various of the well known ideals of B(X). That is, if $\mathcal{I}$ and $\mathcal{q}$ are ideals is $\mathcal{I} \subset \mathcal{q}$? In approaching this problem we introduce the concept of perturbation ideals.

5.2 The ideal of finite dimensional operators F(X)

We begin by proving that the ideal of finite dimensional operators is contained in every nonzero ideal of B(X).

(5.2.1) THEOREM

If I is any nonzero ideal in B(X) and if T is any finite dimensional operator on X, then $T \in I$.

Proof. It suffices to show that if T is a one dimensional operator on X, then $T \in I$, for any finite dimensional operator is the finite sum of such one dimensional operators. Therefore suppose T is one dimensional, then it follows easily that there exist fixed elements $y \in X$ and $x^* \in X^*$ such that

$$T(x) = x^*(x)y \quad \text{for all} \quad x \in X.$$

Choose $W \in I$, $W \neq 0$ and choose $x_o \in X$ such that $W(x_o) \neq 0$. Define the one dimensional operator S by

$$S(x) = x^*(x)x_o \quad \text{for all} \quad x \in X.$$

Choose $x_o^* \in X^*$ such that $x_o^*(y_o) = 1$, where y_o denotes $W(x_o)$. Now define another one dimensional operator V by

$$V(x) = x_o^*(x)y \quad \text{for all} \quad x \in X.$$

Since $W \in I$ we have that $VWS \in I$, since I is a two sided ideal. But $VWS(x) = VW(x^*(x)x_o) = x^*(x)V(y_o) = x^*(x)x_o^*(y_o)y = x^*(x)y = T(x)$ for all $x \in X$. Therefore $T \in I$.■

It is obvious that the set of all finite dimensional operators on X forms an ideal, we denote this ideal by $F(X)$.

(5.2.2) COROLLARY

If I is any nonzero ideal in $B(X)$, then $F(X) \subset I$.

A long outstanding open question has been: What is the closure of $F(X)$? In this direction I. Maddaus [48] proved the following result.

(5.2.3) THEOREM

If X has a basis (see Definition 5.4.1), then $\overline{F(X)} = K(X)$.

Proof. Let S denote the unit ball of X and let $T \in K(X)$. $T(S)$ is therefore totally bounded. If $\{\phi_i\}_{i=1}^{\infty}$ is a basis for X and $x \in X$ with $x = \sum_{i=1}^{\infty} a_i\phi_i$ we define the continuous projection $P_n : X \to X$ by

$$P_n(x) = \sum_{i=1}^{n} a_i\phi_i \quad \text{for each} \quad n = 1,2,\ldots.$$

L. W. Cohen and N. Dunford [13, Theorem 2] proved that $\{P_n(y)\}_{n=1}^{\infty}$ converges uniformly to y for all y in any totally bounded subset of X. Therefore $P_n(T(x))$ converges uniformly to $T(x)$ for all $x \in S$ which implies $||P_nT - T|| \to 0$ as $n \to \infty$. However $P_nT \in F(X)$ for each n and therefore $T \in \overline{F(X)}$.■

(5.2.4) COROLLARY

If X has a basis and I is any nonzero closed two sided ideal of X, then $K(X) \subset I$.

J. Lindenstrauss [47] proved that the spaces $L_p(\mu)$ $(1 \le p \le \infty$, μ an arbitrary positive measure) and $C(K)$, the Banach space of all continuous functions on an extremely disconnected compact Hausdorff space

K, also have the property that $\overline{F(X)} = K(X)$ (for a further discussion see [46, p. 12]).

A recent result of P. Enflo [23] demonstrates an example of a separable reflexive Banach space X where $\overline{F(X)} \neq K(X)$ and where X has no basis.

5.3 The complexity of the ideal structure of B(X)

It must be noted here that a complete classification of all the ideals of B(X) can be an impossible task. Most of the results on ideal structure deal with the well known closed ideals which have arisen from applied work with operators or from natural topological considerations. To name a few, the compact operators, weakly compact operators and strictly singular operators. But the ideal structure of B(X) can be extremely complicated. To illustrate this we present some results due to H. Porta [53].

(5.3.1) DEFINITION

If X and Y are Banach spaces let $m(Y,X)$ denote the set of operators $T \in B(X)$ that can be factored through Y, i.e.

$$m(Y,X) = \{T \in B(X) \mid \exists Q \in B(X,Y) \text{ and } S \in B(Y,X) \ni T = SQ\}.$$

Denote the closure of $m(Y,X)$ by $a(Y,X)$.

(5.3.2) LEMMA

$m(Y,X)$ is a two sided ideal if Y is isomorphic to $Y \times Y$, that is, if there exists a one to one bicontinuous linear transformation from Y onto $Y \times Y$.

Proof. Obviously if $T \in m(Y,X)$ and α is any scalar then

$$\alpha T \in m(Y,X).$$

Let $T_1 = S_1Q_1$ and $T_2 = S_2Q_2$ be elements of $m(Y,X)$ with $S_i \in B(Y,X)$ and $Q_i = B(X,Y)$ for $i = 1,2$.

Define

$$W : X \to Y \times Y \text{ by } W(x) = (Q_1(x), Q_2(x)),$$

$$U_1 : Y \times Y \to X \text{ by } U_1(y_1,y_2) = S_1(y_1),$$

and $$U_2 : Y \times Y \to X \text{ by } U_2(y_1,y_2) = S_2(y_2).$$

Now $T_1 + T_2 = (U_1 + U_2)W$ so $T_1 + T_2 \in m(Y \times Y,X)$. Since Y and $Y \times Y$

are isomorphic this implies $T_1 + T_2 \in m(Y,X)$, therefore $m(Y,X)$ is a subspace of $B(X)$.

Finally, suppose $T \in m(Y,X)$ and $W \in B(X)$. Using the definition it is easy to see that TW and WT are in $m(Y,X)$ without using the fact that Y is isomorphic to $Y \times Y$.■

The following lemma is very useful in what follows.

(5.3.3) LEMMA

$I \in m(Y,X)$ if and only if there exists a Banach space Z such that Y is isomorphic to $X \oplus Z$.

Proof. If $I \in m(Y,X)$ then $I = SQ$, $S \in B(Y,X)$, $Q \in B(X,Y)$, so $P = QS \in B(Y)$ is a continuous projection (i.e. $P^2 = P$). Letting Z and M denote the kernel and range of P, it is clear that Y is isomorphic (in fact equal) to $M \oplus Z$. Also it can easily be seen that Q defines an isomorphism between X and M, and therefore Y is isomorphic to $X \oplus Z$.

Conversely, if $\sigma : Y \to X \oplus Z$ is an isomorphism define

$$Q : X \to Y \text{ by } Q(x) = \sigma^{-1}(x,o)$$

and

$$S : Y \to X \text{ by } S(y) = \text{first coordinate of } \sigma(y) \in X \oplus Z.$$

Then $SQ = I$, so $I \in m(Y,X)$.■

In all that follows, subspace means closed linear subspace.

(5.3.4) LEMMA

Let X be a Banach space and Y a complemented subspace of X. Then, for an arbitrary space Z, the following conditions are equivalent:

(i) $m(Y,X) \subset a(Z,X)$

(ii) Y is isomorphic to a complemented subspace of Z.

Proof. Let $P \in B(X)$ be a projection onto Y, $I : Y \to Y$ the identity and $J : Y \to X$ the canonical injection; it is clear that $P \in m(Y,X)$. Let $\varepsilon > 0$ be such that $\varepsilon \, ||P|| < 1$.

First suppose that $m(Y,X) \subset a(Z,X)$. Then there exist $S \in B(Z,X)$ and $Q \in B(X,Z)$ such that $||P - SQ|| < \varepsilon$. Consider the operator $U \in B(Y)$ defined by $U = I - PSQJ$: since $I = PJ$ we see that

$U = PJ - PSQJ = P(P - SQ)J$ and therefore

$$||U|| \leq ||P|| \, ||P - SQ|| \, ||J|| \leq ||P||\cdot\varepsilon < 1.$$

Therefore $PSQJ \in B(Y)$ is invertible, that is, there exists $T \in B(Y)$ such that $I = TPSQJ = (TPS)(QJ)$. But $TPS \in B(Z,Y)$ and $QJ \in B(Y,Z)$ implies $I \in m(Q,Y)$, so by Lemma 5.3.3 we conclude that Y is isomorphic to a complemented subspace of Z, as desired.

The converse is obvious: if Y' is a complemented subspace of Z isomorphic to Y, then

$$m(Y,X) = m(Y',X) \subset m(Q,X) \subset a(Z,X).\blacksquare$$

(5.3.5) LEMMA

Assume that $X, Y_1, Y_2, \ldots, Y_n$ are Banach spaces such that Y_j is isomorphic to $Y_j \times Y_j$ for $j = 1,2,\ldots,n$. Then

$$m(Y_1,X) + m(Y_2,X) + \cdots + m(Y_n,X) = m(Y_1 \times Y_2 \times \cdots \times Y_n,X).$$

Proof. An inductive argument reduces the proof to the case $n = 2$, which is disposed of as follows. Since Y_1 and Y_2 are (isomorphic to) complemented subspaces of $Y_1 \times Y_2$, it is clear that

$m(Y_1,X) \subset m(Y_1 \times Y_2,X)$ and $m(Y_2,X) \subset m(Y_1 \times Y_2,X)$, whence

$$m(Y_1,X) + m(Y_2,X) \subset m(Y_1 \times Y_2,X).$$

Conversely, if $T = SQ \in m(Y_1 \times Y_2,X)$, where $S \in B(Y_1 \times Y_2,X)$ and $Q \in B(X,Y_1 \times Y_2)$ with $Q(x) = (Q_1(x),Q_2(x))$, then we define $S_1 \in B(Y_1,X)$ and $S_2 \in B(Y_2,X)$ as

$$S_1(y) = S(y,o) \quad \text{and} \quad S_2(y) = S(o,y);$$

finally, let $T_1,T_2 \in B(X)$ be the operators $T_1 = S_1Q_1$ and $T_2 = S_2Q_2$. Clearly $T_1 + T_2 = T$ with $T_j \in m(Y_j,X)$ for $j = 1,2$, and therefore $T \in m(Y_1,X) + m(Y_2,X)$; the lemma follows.$\blacksquare$

From Lemma 5.3.3 and [2, Theorem 7], we obtain that for $p \neq q$, $p \geqq 1$, $q \geqq 1$, the ideal $m(\ell_q,\ell_p)$ is not the whole of $B(\ell_p)$. Since by Corollary 5.4.9 the ideal $K(\ell_p)$ of compact operators is the largest proper closed two sided ideal of $B(\ell_p)$ it follows that

(5.3.6) LEMMA

If $p, q \geqq 1, p \neq q$, then $m(\ell_q, \ell_p) \subset K(\ell_p)$.

The following theorem demonstrates how complicated the ideal structure of $B(X)$ can be.

(5.3.7) THEOREM

There exists a Banach space X with the properties:

(i) X is separable, isometrically isomorphic to its dual X^*, and reflexive;

(ii) it is possible to assign a closed two sided ideal $a(F) \subset B(X)$ to each finite set of positive integers F, in such a way that the mapping $F \to a(F)$ is one to one and inclusion preserving in both directions: $F \subset G$ if and only if $a(F) \subset a(G)$.

Proof. Let P be a countable set of integers $p \geq 1$; define Y the product $Y = \pi\{\ell_p : p \in P\}$. We denote by $|x|$ the norm of $x \in \ell_p$, for all p. Consider the set $\ell(P)$ of all families $\{x_p \in \ell_p : p \in P\} \in Y$ such that $\Sigma\{|x_p|^2 : p \in P\} < \infty$ (this is always the case, if P is finite). It can be seen that $\ell(P)$ is a linear subspace of Y and that the norm $||\{x_p\}|| = (\Sigma|x_p|^2)^{\frac{1}{2}}$ makes $\ell(P)$ a separable Banach space; if P is finite, $\ell(P) = \pi\{\ell_p : p \in P\}$. It is clear that for each subset $Q \subset P$, the space $\ell(Q)$ can be identified with a complemented subspace of $\ell(P)$. Moreover, $\ell(P)$ is always isomorphic to its square $\ell(P) \times \ell(P)$. The dual $(\ell(P))^*$ of $\ell(P)$ can be identified with $\ell(P^*)$, where P^* is the set of conjugates p^* of elements $p \in P$, i.e. $\frac{1}{p} + \frac{1}{p^*} = 1$. In particular, if $1 \notin P$, then $\ell(P)$ is reflexive, and furthermore, if $P = P^*$, then $\ell(P)$ is isometrically isomorphic to its dual. Therefore such $\ell(P)$ satisfy condition (i) above.

Let P be a fixed countably infinite set of integers $p > 1$ such that $P = P^*$, and let X denote the space $X = \ell(P)$. For each finite subset $F \subset P$, let $a(F) \subset B(X)$ be the ideal $a(F) = a(\ell(F), X)$. Since $\ell(F)$ is (isomorphic to) a complemented subspace of $\ell(G)$, whenever $F \subset G$ it is clear (Lemma 5.3.4) that the mapping $F \to a(F)$ is inclusion preserving. On the other hand, suppose that $a(F) \subset a(G)$, or, equivalently, $m(\ell(F),X) \subset a(\ell(G),X)$. By Lemma 5.3.5, this inequality is equivalent to

$$m(\ell_p, X) \subset a(\ell(G), X), \quad \text{for all } p \in F.$$

Lemma 5.3.4 applies, and we conclude that for $p \in F$, ℓ_p is isomorphic to a complemented subspace of $\ell(G)$. By Lemma 5.3.3 this amounts to

$$m(\ell(G),\ell_p) = B(\ell_p).$$

But, again from Lemma 5.3.5,

$$m(\ell(G),\ell_p) = \Sigma\{m(\ell_q,\ell_p) : q \in G\}.$$

Now, if $p \notin G$, from Lemma 5.3.6 it follows that $m(\ell_q,\ell_p) \subset K(\ell_p)$ for all $q \in G$, or $B(\ell_p) = m(\ell(G),\ell_p) \subset K(\ell_p)$, which is absurd. So $p \in G$ for all $p \in F$, and this means $F \subset G$. Therefore we have shown that $F \subset G$ if and only if $a(F) \subset a(G)$, which implies that $F \to a(F)$ is one to one, and the theorem is proved.■

5.4 Uniqueness of K(X)

The example of section 5.3 demonstrates how complex the ideal structure of B(X) can be, on the other hand, J. W. Calkin [7] showed the fascinating result that, for $X = \ell_2$, K(X) is the unique nonzero proper closed two sided ideal in B(X). A very interesting extension of this result was discovered by I. Gohberg, A. Markus and I. Fel'dman [28], who obtained the same conclusion for $X = \ell_p$, $1 \leq p < \infty$ and $X = c_o$. This work received further treatment at the hands of R. H. Herman [32] whose exposition we incorporate here.

The preliminary material on bases goes back to Schauder [60].

(5.4.1) DEFINITION

Let X be an infinite dimensional Banach space. A sequence $\{\Phi_i\}$ in X is called a basis for X if given any $x \in X$ there exists a unique sequence of scalars $\{a_i\}$ such that $x = \sum_{i=1}^{\infty} a_i\Phi_i$, that is $\lim_{n\to\infty} ||x - \sum_{i=1}^{n} a_i\Phi_i|| = 0$. Note that $\Phi_i \neq 0$ is true for all i.

If $\{\Phi_i\}$ is a basis for X and $||\Phi_i|| = 1$ for all i, then we call $\{\Phi_i\}$ a normalized basis for X.

We define the sequence of coefficient functionals $\{g_i\}$ associated with $\{\Phi_i\}$ by

$$g_i(x) = a_i \quad \text{for each } x \in X, \text{ where } x = \sum_{i=1}^{\infty} a_i\Phi_i.$$

By the fact that the coefficients for each $x \in X$ are unique it follows that each of the functionals g_i is linear. We wish to show that each g_i is continuous for any basis $\{\Phi_i\}$ and that the norms of the g_i are uniformly bounded if $\{\Phi_i\}$ is a normalized basis.

(5.4.2) LEMMA

Let $\{x_i\}$ be a fixed sequence in a Banach space X, such that $x_i \neq 0$ for $i = 1,2,\ldots$ and let A be the linear space of sequences of scalars

$$A = \{\{\alpha_i\} \mid \sum_{i=1}^{\infty} \alpha_i x_i \text{ converges in } X\}$$

endowed with the norm

$$||\{\alpha_i\}|| = \sup_{1 \leq n < \infty} ||\sum_{i=1}^{n} \alpha_i x_i||. \tag{5.4.3}$$

Then A is a Banach space.

Proof. As defined the number $||\{\alpha_i\}||'$ is finite, since the sequence $\{||\sum_{i=1}^{n} \alpha_i x_i||\}$ is convergent. Since all $x_i \neq 0$, equation (5.4.3) defines a norm on the linear space A.

Let $\{\alpha_i^{(k)}\}_{k=1}^{\infty}$ be a Cauchy sequence in A. Then for each $\varepsilon > 0$ there exists a positive integer N_ε such that

$$||\{\alpha_i^{(k)}\} - \{\alpha_i^{(m)}\}|| = \sup_{1 \leq n < \infty} ||\sum_{i=1}^{n} (\alpha_i^{(k)} - \alpha_i^{(m)})x_i|| < \varepsilon \quad \text{for } m,k > N_\varepsilon.$$

Hence for each $n = 1,2,\ldots$ we have

$$||(\alpha_n^{(k)} - \alpha_n^{(m)})x_n|| \leq ||\sum_{i=1}^{n} (\alpha_i^{(k)} - \alpha_i^{(m)})x_i|| + ||\sum_{i=1}^{n-1} (\alpha_i^{(k)} - \alpha_i^{(m)})x_i|| < 2\varepsilon$$

for $m,k > N_\varepsilon$, therefore since all the $x_n \neq 0$ it follows that

$$|\alpha_n^{(k)} - \alpha_n^{(m)}| < \frac{2\varepsilon}{||x_n||} \quad \text{for } m,k > N_\varepsilon \text{ and for all } n.$$

Consequently for fixed $n \geq 1$ the sequence of scalars $\{\alpha_n^{(k)}\}_{k=1}^{\infty}$ is convergent to a scalar α_n. Hence, from the inequality

$||\sum_{i=1}^{n} (\alpha_i^{(k)} - \alpha_i^{(m)})x_i|| < \varepsilon$ for $m,k > N_\varepsilon$ we obtain by letting $m \to \infty$ that

$$||\sum_{i=1}^{n} (\alpha_i^{(k)} - \alpha_i)x_i|| \leq \varepsilon \quad \text{for } k > N_\varepsilon.$$

Then $||\sum_{i=n+1}^{n+p} \alpha_i x_i|| \leq 2\varepsilon + ||\sum_{i=n+1}^{n+p} \alpha_i^{(k)} x_i||$ for $k > N_\varepsilon$ and $n,p=1,2,\ldots$.

Therefore, since each series $\sum_{i=1}^{\infty} \alpha_i^{(k)} x_i$ is convergent and since X is complete, it follows that $\sum_{i=1}^{\infty} \alpha_i x_i$ converges in X, i.e. $\{\alpha_i\} \in A$. Moreover, by the above we have

$$||\{\alpha_i^{(k)}\} - \{\alpha_i\}|| = \sup_{1 \leq n < \infty} ||\sum_{i=1}^{n} (\alpha_i^{(k)} - \alpha_i)x_i|| \leq \varepsilon \quad \text{for } k > N_\varepsilon,$$

which completes the proof.■

(5.4.4) LEMMA

Let X be a Banach space with basis $\{\Phi_i\}$. Let A be the Banach space defined from the sequence $\{\Phi_i\}$ using Lemma 5.4.2. Then the mapping

$$\tau : \{a_i\} \to \sum_{i=1}^{\infty} a_i \Phi_i$$

is an isomorphism from A onto X. (That is τ is a bicontinuous, one to one linear transformation from A onto X).

Proof. Since $\Phi_i \neq 0$ for all i, Lemma 5.4.2 may be applied. The mapping τ from A into X is obviously linear and of norm 1. Since $\{\Phi_i\}$ is a basis for X it follows that τ is one to one because of the uniqueness and existence of the expansions $x = \sum_{i=1}^{\infty} a_i \Phi_i$. Hence, by the inversion theorem of Banach [22], τ is an isomorphism from A onto X.■

(5.4.5) THEOREM

Let X be a Banach space with basis $\{\Phi_i\}$. Then the coefficient functionals g_i associated with $\{\Phi_i\}$ are continuous, i.e. $g_i \in X^*$ for $i = 1,2,\ldots$. Moreover, if $\{\Phi_i\}$ is a normalized basis then there exists a constant M such that

$$||g_i|| \leq M \text{ for } i = 1,2,\ldots.$$

Proof. Using Lemma 5.4.4 we get, for each $x \in X$ and $i = 1,2,\ldots,$ the inequalities

$$|g_i(x)| = \frac{||g_i(x)\Phi_i||}{||\Phi_i||} = \frac{1}{||\Phi_i||} \left|\left| \sum_{j=1}^{i} g_j(x)\Phi_j - \sum_{j=1}^{i-1} g_j(x)\Phi_j \right|\right|$$

$$\leq \frac{1}{||\Phi_i||} 2||\{g_i(x)\}_{i-1}^{\infty}||_A = \frac{2}{||\Phi_i||} ||\tau^{-1}(x)||_A$$

$$\leq \frac{2}{||\Phi_i||} ||\tau^{-1}|| \, ||x||.$$ Therefore g_i is continuous.

Moreover if $\{\Phi_i\}$ is a normalized basis then, since $||\Phi_i|| = 1$,

$$||g_i|| \leq 2||\tau^{-1}|| = M,$$

for $i = 1,2,\ldots.$ ■

(5.4.6) DEFINITION

Let $\{\Phi_i\}$ be a basis for X. $\{z_k\}$ is said to be a block basis if for a fixed sequence of nonnegative integers $0 = a_1 < a_2 < \ldots$ we have

$$z_k = \sum_{i=a_k+1}^{a_{k+1}} b_i \Phi_i$$

where $z_k \neq 0$ and the $\{b_i\}$ are some fixed scalars.

(5.4.7) THEOREM

Let X be a Banach space. A sequence $\{x_n\}$ in X with $x_n \neq 0$ is a basis for $\overline{\text{span}}\{x_n\}$ if and only if there exists a constant $k \geq 1$ such that the inequality

(5.4.8)

$$||t_1x_1 + t_2x_2 + \cdots + t_px_p|| \leq k||t_1x_1 + \cdots + t_px_p + t_{p+1}x_{p+1} + \cdots + t_qx_q||$$

is satisfied for all arbitrary positive integers $p \leq q$ and for all arbitrary scalars $t_1,t_2,\ldots,t_q$. ($\overline{\text{span}\{x_n\}}$ denotes closure in X of the linear span of $\{x_n\}$).

Proof. ($\rightarrow$) Suppose $\{x_n\}$ is a basis for $\overline{\text{span}\{x_n\}}$. Let $p \leq q$ be positive integers and $t_1,t_2,\ldots,t_q$ be any scalars.

If $x = \sum_{j=1}^{\infty} a_jx_j \in \overline{\text{span}\{x_n\}}$, then by Lemma 5.4.4 we have that there exists a $k \geq 1$ (independent of x) such that

$||\{a_j\}||_A = \sup_n||\sum_{j=1}^{n} a_jx_j|| \leq k||x||$. Therefore if $x = \sum_{j=1}^{q} t_jx_j$ we have

$||\sum_{j=1}^{p} t_jx_j|| \leq k||x|| = k||\sum_{j=1}^{q} t_jx_j||$, which is inequality (5.4.8).

($\leftarrow$) Suppose $\{x_n\}$, $x_n \neq 0$ satisfies inequality (5.4.8). Consider the sequence space A of Lemma 5.4.2 corresponding to $\{x_n\}$. Suppose $x \in \overline{\text{span}\{x_n\}}$. Then $x = \lim_{i\to\infty} y_i$ where each y_i is a finite linear combination of the x_n's, that is $y_i = \sum_{j=1}^{m_i} \alpha_j^{(i)}x_j$. Set $\alpha_j^{(i)} = 0$ if $j > m_i$ and note that without loss of generality we may assume that $m_1 \leq m_2 \leq m_3 \leq \cdots$. The sequence of scalars $w_i = \{\alpha_1^{(i)},\alpha_2^{(i)},\ldots\}$ lies in A for $i = 1,2,\ldots$. Now for $p > q$ we have that

$$||w_p - w_q||_A = \sup_r||\sum_{j=1}^{r} (\alpha_j^{(p)} - \alpha_j^{(q)})x_j|| \leq k \sum_{j=1}^{m_p} (\alpha_j^{(p)} - \alpha_j^{(q)})x_j||$$

$= k||y_p - y_q||$. Therefore since $\{y_i\}$ is a Cauchy sequence in X, we have that $\{w_i\}$ is a Cauchy sequence in A. By the proof of Lemma 5.4.2 we now know that w_i converges to w in A where $w = \{\alpha_j\}$ and $\alpha_j = \lim_{i\to\infty} \alpha_j^{(i)}$. We then know that $y = \sum_{j=1}^{\infty} \alpha_jx_j \in X$, by the definition of A. We now show that y_i converges to y in X.

$||w_i - w||_A = \sup_r || \sum_{j=1}^{r} (\alpha_j^{(i)} - \alpha_j)x_j||$, but $\alpha_j^{(i)} = 0$ for all $i > m_i$,

therefore $||w_i - w||_A \geq ||y_i - \sum_{j=1}^{r} \alpha_j x_j||$ for all $r > m_i$, so letting

$r \to \infty$ we get $||w_i - w||_A \geq ||y_i - y||$ and since w_i converges to w we get that y_i converges to y. We have shown that $y_i \to x$ and $y_i \to y$, so $x = y = \sum_{j=1}^{\infty} \alpha_j x_j$. To complete the proof that $\{x_n\}$ is a basis we must show that the coefficients $\{\alpha_j\}$ are unique for x. Suppose

$$x = \sum_{j=1}^{\infty} \alpha_j x_j \quad \text{and} \quad x = \sum_{j=1}^{\infty} \beta_j x_j.$$

then

$$0 = \sum_{j=1}^{\infty} (\alpha_j - \beta_j)x_j.$$

But suppose $a_j - b_j \neq 0$ for some j and let j be the first positive integer for which this occurs. Then

$0 = ||0|| = || \sum_{i=j}^{\infty} (\alpha_i - \beta_i)x_i|| = \lim_{n\to\infty} || \sum_{i=j}^{n} (\alpha_i - \beta_i)x_i||$. But for each

$n > j$, $0 < \varepsilon = ||(\alpha_j - \beta_j)x_j|| \leq k|| \sum_{i=j}^{n} (\alpha_i - \beta_i)x_i||$ by hypothesis

(inequality (5.4.8)). Hence $\lim_{n\to\infty} || \sum_{i=j}^{n} (\alpha_i - \beta_i)x_j|| \geq \frac{\varepsilon}{k} > 0$, a contradiction. Therefore $\alpha_i = \beta_i$ for $i = 1,2,\ldots,$ proving that $\{x_n\}$ is a basis for $\overline{\text{span}}\{x_n\}$.■

With the characterization of Theorem 5.4.7 it is clear that any block basis $\{z_k\}$ is a basis for $\overline{\text{span}}\{z_k\}$.

(5.4.9) LEMMA

Let $\{x_n\}$ be a basis for a Banach space X where $\{g_i\}$ are the coefficient functionals. Suppose k is the constant of Lemma 5.4.7 for this basis. If $\inf||x_n|| = c > 0$ then each $||g_i|| \leq 2kc^{-1}$.

Proof. From 5.4.8 we see that $\sup_n \|\sum_{j=1}^{n} g_j(x)x_j\| \leqq \|x\| = \|\sum_{j=1}^{\infty} g_j(x)x_j\|$,

so $|g_i(x)|\ \|x_i\| \leqq \|\sum_{j=1}^{i} g_j(x)x_j\| + \|\sum_{j=1}^{i-1} g_j(x)x_j\| \leqq 2k\|x\|$.

From this we get $\|g_i\| \leqq 2kc^{-1}$.■

(5.4.10) LEMMA

Let $T \in B(X)$, X a Banach space with a normalized basis $\{\Phi_i\}$. A sufficient condition for T to be compact is that

$$\sum_{k=1}^{\infty} \|T(\Phi_k)\| < \infty.$$

Proof. It suffices to show that if S is the unit ball of X, then $T(S)$ is totally bounded. Let g_j denote the jth coefficient functional. By Theorem 5.4.5 the g_j are continuous, in fact there exists M such that $\|g_j\| < M$ for all j.

For each $x = \sum_{i=1}^{\infty} a_i\Phi_i \in S$ define $S_N(x) = \sum_{i=1}^{N} a_i\Phi_i$ and

$P_N(x) = \sum_{i=N+1}^{\infty} a_i\Phi_i$ for $N = 1,2,\ldots$. Let $\varepsilon > 0$ be given. Choose a

positive integer N such that $\sum_{i=N+1}^{\infty} \|T(\Phi_i)\| < \frac{\varepsilon}{3M}$, then for all

$x = \sum_{i=1}^{\infty} a_i\Phi_i \in S$ we have

$$\|T(P_N(x))\| = \|\sum_{i=N+1}^{\infty} a_iT(\Phi_i)\| \leq \sum_{i=N+1}^{\infty} |a_i|\ \|T(\Phi_i)\| \leq M \sum_{i=N+1}^{\infty} \|T(\Phi_i)\|$$

$$< M\frac{\varepsilon}{3M} = \frac{\varepsilon}{3}, \text{ since}$$

$$|a_j| = |g_j(x)| \leq \|g_j\|\ \|x\| \leq \|g_j\| < M \text{ for all } j.$$

Let $Y = \{T(S_N(x)) | x \in S\}$. Clearly Y is a bounded subset of the finite dimensional space $\text{span}[T(\Phi_1),T(\Phi_2),\ldots,T(\Phi_N)]$, so Y is totally bounded in $T(S)$. Therefore there exist $z_1,z_2,\ldots,z_j$ in S such that

$$||T(S_N(y)) - T(S_N(z_k))|| < \frac{\varepsilon}{3} .$$

Therefore

$$||T(y) - T(z_k)|| = ||T(S_N(y)) - T(S_N(z_k)) + T(P_N(y)) - T(P_N(z_k))||$$

$$\leq ||T(S_N(y)) - T(S_N(z_k))|| + ||T(P_N(y))|| + ||T(P_N(z_k))|| < \frac{\varepsilon}{3} + \frac{\varepsilon}{3} + \frac{\varepsilon}{3} = \varepsilon.$$

Therefore $T(S)$ is totally bounded.■

The following results on bases are due to C. Bessaga and A. Pełczyński [4].

(5.4.11) DEFINITION

Let $\{x_n\}$ be a basis for a Banach space X and $\{y_n\}$ be a basis for a Banach space Y. We say that the bases $\{x_n\}$ and $\{y_n\}$ are equivalent if $\sum_{i=1}^{\infty} t_i x_i$ converges in X if and only if $\sum_{i=1}^{\infty} t_i y_i$ converges in Y. In this case, the mapping $T(\Sigma t_k x_k) = \Sigma t_k y_k$ is an isomorphism between X and Y, that is, T is a bicontinuous linear transformation of X onto Y, by the closed graph theorem [22, p. 57].

(5.4.12) LEMMA

Let $\{x_n\}$ be a basis for the Banach space X. If the sequence $\{y_n\} \subset X$ satisfies the condition

(5.4.13) $$\sum_{n=1}^{\infty} ||x_n - y_n|| \, ||g_n|| = \delta < 1$$

where the g_n are the coefficient functionals for $\{x_n\}$, then $\{x_n\}$ and $\{y_n\}$ are equivalent.

Proof. Since

$$|t_i| = |g_i(t_1 x_1 + \cdots + t_p x_p)| \leq ||g_i|| \, ||t_1 x_1 + \cdots + t_p x_p|| \quad \text{for } i \leq p$$

we obtain

$$||t_1y_1 + \cdots + t_py_p|| \leq ||t_1x_1 + \cdots + t_px_p|| + \sum_{i=1}^{p} |t_i|\ ||x_i - y_i||$$

$$\leq ||t_1x_1 + \cdots + t_px_p|| + \sum_{i=1}^{p} ||t_1x_1 + \cdots + t_px_p||\ ||g_i||\ ||x_i - y_i||$$

$$\leq (1 + \delta)||t_1x_1 + \cdots + t_px_p||,$$

and

$$||t_1y_1 + \cdots + t_qy_q|| \geq ||t_1x_1 + \cdots + t_qx_q|| - \sum_{i=1}^{q} |t_i|\ ||x_i - y_i||$$

$\geq (1 - \delta)||t_1x_1 + \cdots + t_qx_q||$ for arbitrary positive integers $p \leq q$ and for arbitrary scalars $t_1, t_2, \ldots, t_q$.

Therefore

$$||t_1y_1 + \cdots + t_py_p|| \leq ||t_1y_1 + \cdots + t_qy_q||\ k\, \frac{1 + \delta}{1 - \delta},$$

where k is the constant from Theorem 5.4.7 for the basis $\{x_n\}$. Therefore by Theorem 5.4.7, $\{y_n\}$ is a basis for $\overline{\text{span}\{y_n\}}$. The equivalence of $\{x_n\}$ and $\{y_n\}$ is an immediate consequence of our inequalities.∎

(5.4.14) LEMMA

Let $\{x_n\}$ be a basis for a Banach space X. If a sequence $\{y_n\} \subset X$ satisfies the conditions

$$\inf_n ||y_n|| = \varepsilon > 0$$

(5.4.15)

$$\lim_{n\to\infty} g_i(y_n) = 0$$

for all the coefficient functionals g_i of $\{x_n\}$, then there exists a subsequence $\{y_{n_k}\}$ which is a basis for the closure of its span and this basis $\{y_{n_k}\}$ is equivalent to a block basis (with respect to $\{x_n\}$).

Proof. We use several times the fact that, for each $x \in X$

$$x = \sum_{i=1}^{\infty} g_i(x)x_i.$$

Let k be the constant of Lemma 5.4.7 for the basis $\{x_i\}$. By induction we choose increasing sequences $\{p_n\}$ and $\{q_n\}$ of positive integers as follows.

Let $p_1 = 1$. Choose q_1 so large that

$$\frac{4k}{\varepsilon} \left|\left| \sum_{i=q_1+1}^{\infty} g_i(y_{p_1})x_i \right|\right| < \frac{1}{2^{1+2}} .$$

Then choose by (5.4.15) $p_2 > p_1$ such that

$$\frac{4k}{\varepsilon} \left|\left| \sum_{i=1}^{q_1} g_i(y_{p_2})x_i \right|\right| < \frac{1}{2^{1+2}} .$$

By induction if $p_1, \ldots, p_n$ and $q_1, \ldots, q_{n-1}$ have been selected we pick $q_n > q_{n-1}$ and $p_{n+1} > p_n$ so that

$$\frac{4k}{\varepsilon} \left|\left| \sum_{i=q_n+1}^{\infty} g_i(y_{p_n})x_i \right|\right| < \frac{1}{2^{n+2}} .$$

$$\frac{4k}{\varepsilon} \left|\left| \sum_{i=1}^{q_n} g_i(y_{p_{n+1}})x_i \right|\right| < \frac{1}{2^{n+2}} .$$

Set $z_n = \sum_{i=q_n+1}^{q_{n+1}} g_i(y_{p_{n+1}})x_i$ and note that z_n is part of the expansion of $y_{p_{n+1}}$ in terms of the basis x_i.

Now

$$||z_n|| \leqq \left|\left| \sum_{i=1}^{q_{n+1}} g_i(y_{p_{n+1}})x_i \right|\right| + \left|\left| \sum_{i=1}^{q_n} g_i(y_{p_{n+1}})x_i \right|\right|$$

$$\leqq 2k||y_{p_{n+1}}|| \quad \text{(see the proof of 5.4.9).}$$

Also

$$||z_n|| \geqq ||y_{p_{n+1}}|| - \left|\left| \sum_{i=1}^{q_n} g_i(y_{p_{n+1}})x_i \right|\right| - \left|\left| \sum_{i=q_{n+1}+1}^{\infty} g_i(y_{p_{n+1}})x_i \right|\right| \geqq \varepsilon/2$$

by the definition of p_{n+1}, q_n, g_{n+1} and by 5.4.15.

Also

$$\frac{4k}{\varepsilon}\,||y_{p_{n+1}} - z_n||$$

$$\leq \frac{4k}{\varepsilon}\left[||\sum_{i=1}^{q_n} g_i(y_{p_{n+1}})x_i|| + ||\sum_{i=q_{n+1}+1}^{\infty} g_i(y_{p_{n+1}})x_i||\right]$$

$$\leq \frac{1}{2^{n+2}} + \frac{1}{2^{n+3}} < \frac{1}{2^{n+1}}\ .$$

Summarizing we have

$$\varepsilon/2 \leq ||z_n|| \leq 2k||y_{p_{n+1}}||,\quad n = 1,2,\ldots \quad \text{and}$$

(5.4.16)
$$\sum_{n=1}^{\infty} \frac{4k}{\varepsilon}\,||y_{p_{n+1}} - z_n|| < \frac{1}{2}\ .$$

Let $\{h_n\}$ be the coefficient functionals for the block basis $\{z_n\}$. Note that k is the constant of Lemma 5.4.7 for that basis (as well as for the basis $\{x_n\}$). Therefore Lemma 5.4.9 shows that

$$||h_n|| \leq 4k\varepsilon^{-1}.$$

Applying Lemma 5.4.12 we see that the sequence $\{y_{p_{n+1}}\}$ is a basis (for the closure of its span) and is equivalent to the block basis $\{z_n\}$.∎

(5.4.17) DEFINITION

Let X be a Banach space with a normalized basis $\{\Phi_i\}$. We call $\{\Phi_i\}$ perfectly homogeneous if $\{\Phi_i\}$ is equivalent to every normalized block basis $\{z_n\}$ (with respect to $\{\Phi_i\}$).

(5.4.18) DEFINITION

Let X be a Banach space with a normalized basis $\{\Phi_i\}$. We say that $\{\Phi_i\}$ has (+) if given any normalized block basis $\{z_k\}$, with respect to $\{\Phi_i\}$, there exists $P : X \to \overline{\text{span}}\{z_k\}$, where P is a continuous projection onto $\overline{\text{span}}\{z_k\}$.

(5.4.19) LEMMA

Let X be a Banach space with normalized basis $\{\Phi_n\}$. If $\{\Phi_n\}$ is perfectly homogeneous and has $(+)$, then no proper closed ideal $\Omega \subset B(X)$ can contain an operator U such that

$$\inf_n ||U(\Phi_n)|| = \varepsilon > 0 \quad \text{and} \quad \lim_{n\to\infty} g_i(U(\Phi_n)) = 0$$

for each coordinate functional g_i with respect to $\{\Phi_n\}$.

Proof. Suppose such an operator U was in Ω. We use the notation of Lemma 5.4.14 with $y_n = U(\Phi_n)$. By Lemma 5.4.14 it follows that some subsequence $\{U(\Phi_{p_{n+1}})\}$ is a basis and is equivalent to a block basis $\{z_n\}$ with $\frac{\varepsilon}{2} \leq ||z_n|| \leq 2k||U(\Phi_{p_{n+1}})|| \leq 2k||U||$. Therefore the normalized block basis $\tilde{z}_n = \dfrac{z_n}{||z_n||}$ is also equivalent to $\{U(\Phi_{p_{n+1}})\}$. But $\{\tilde{z}_n\}$ by hypothesis is equivalent to $\{\Phi_n\}$. Therefore $\{\Phi_n\}$ is equivalent to $\{z_n\}$ which is equivalent to $\{U(\Phi_{p_{n+1}})\}$. Hence there exist T_1 and $T_2 \in B(X)$ such that $T_2(\Phi_n) = U(\Phi_{p_{n+1}})$ and $T_1(\Phi_n) = z_n$ for $n = 1,2,\ldots$. Therefore $\sum_{n=1}^{\infty} ||(T_1 - UT_2)\Phi_n|| = \sum_{n=1}^{\infty} ||z_n - U(\Phi_{p_{n+1}})|| < \infty$, by (5.4.16). We now apply Lemma 5.4.10 to get the $T_1 - UT_2 \in K(X)$. By assumption $\{\Phi_n\}$ has $(+)$, therefore, there exists a continuous projection $P_1 : X \to \overline{\text{span}}\{\tilde{z}_n\}$ and hence a continuous projection $P : X \to \overline{\text{span}}\{z_n\}$.

By Corollary 5.2.4, $T_1 - UT_2 \in \Omega$. Therefore, $T_1^{-1} P(T_1 - UT_2) = I - T_1^{-1} P U T_2$ and since $U \in \Omega$, then $I \in \Omega$ which implies $\Omega = B(X)$, a contradiction.■

(5.4.20) THEOREM

Let X be as in Lemma 5.4.19. If Ω is a nonzero proper closed ideal in $B(X)$, then Ω is the ideal of compact operators.

Proof. Let A be a non-compact operator and $A \in \Omega$. Then there exists a sequence $\{x_n\}$, $||x_n|| = 1$, such that $\{A(x_n)\}$ has no convergent subsequence. If g_j denotes the coefficient functionals for $\{\Phi_n\}$, we may, by the diagonalization process, extract a subsequence $\{A(x_{n_k})\}$ of $\{A(x_n)\}$ such that $g_j A(x_{n_k})$ converges for each j. Note that $\{A(x_{n_k})\}$ has no convergent subsequence so we may choose a further subsequence $\{\bar{x}_\ell\}$ such that $\inf_\ell ||A(\bar{x}_\ell) - A(\bar{x}_{\ell+1})|| = \varepsilon > 0$. Let $z_\ell = \bar{x}_\ell - \bar{x}_{\ell+1}$; then

(1) $\lim_{\ell\to\infty} g_j(A(z_\ell)) = 0$ for each j and $\inf_\ell ||A(z_\ell)|| = \varepsilon > 0$.

By the diagonalization process, extract a subsequence of $\{z_\ell\}$ such that $g_j(z_{\ell_k})$ converges for each j. Now $\{A(z_{\ell_k})\}$ has no convergent subsequence for, if it did, say $A(z_r) \to x$, then $g_j(x) = 0$ for all j by the continuity of g_j, hence $x = 0$. But $\inf_r ||A(z_r)|| = \varepsilon > 0$, therefore $x \neq 0$, a contradiction. Therefore, there exists a subsequence $\{\bar{z}_r\}$ of $\{z_{\ell_k}\}$ such that

(2) $\inf_r ||A(\bar{z}_r - \bar{z}_{r+1})|| = b > 0$.

Let $y_r = \bar{z}_r - \bar{z}_{r+1}$. Now $g_j(y_r) \to 0$ for each j and by (2) there exists $\delta > 0$ such that $\inf_r ||y_r|| = \delta$. By Lemma 5.4.14 there is a subsequence $\{y_{p_{n+1}}\}$ of $\{y_r\}$ which is a basis and is equivalent to a block basis $\{z_n\}$ where $\frac{\delta}{2} \leq ||z_n|| \leq 2k||y_{p_{n+1}}||$ for each n. Clearly $\{y_r\}$ is a bounded sequence, so $\{z_n\}$ is equivalent to the normalized block basis $\left\{\frac{z_n}{||z_n||}\right\}$ which, by hypothesis, is equivalent to $\{\Phi_n\}$. We conclude that $\{\Phi_n\}$ is equivalent to $\{y_{p_{n+1}}\}$. Therefore there exists $T \in B(X)$ such that $T(\Phi_n) = y_{p_{n+1}}$. Then

$$g_j(AT\Phi_n) = g_j A(y_{p_{n+1}}) \to 0 \text{ by (1) since } y_{p_{n+1}} = \bar{z}_{p_{n+1}} - \bar{z}_{(p_{n+1}+1)}.$$

Also

$$\inf_n ||AT(\Phi_n)|| \geqq b > 0 \text{ by (2).}$$

Lemma 5.4.19 asserts that AT lies in no proper closed ideal. But $AT \in \Omega$, a contradiction. Therefore $\Omega \subset K(X)$ and using Corollary 5.2.4 we get that $\Omega = K(X)$.■

(5.4.21) LEMMA

The usual normalized bases in the spaces ℓ_p $(1 \leq p < \infty)$ and c_o are perfectly homogeneous.

Proof. Easy to verify.■

(5.4.22) LEMMA

The usual normalized bases in the spaces ℓ_p $(1 \leq p < \infty)$ and c_o have (+).

Proof. Let $\{z_n\}$ be a normalized block basis in c_o. Let h_i be the corresponding coefficient functionals. By Lemma 5.4.9, $||h_i|| \leq 2$ for all i. By the Hahn-Banach theorem we extend each h_i to h_i' a continuous linear functional on c_o. Hence, h_k' belongs to ℓ_1 for all k,

Suppose $z_k = \sum_{a_k+1}^{a_{k+1}} b_i e_i$, where $\{e_i\}$ is the standard basis for c_o. Each

$$h_k' = \sum_{i=1}^{\infty} a_i^{(k)} f_i$$

and $||h_k'|| \leq 2$ ($\{f_i\}$ is the standard basis in ℓ_1). By the form of the norm in ℓ_1 we may take

$$h_k'' = \sum_{a_k+1}^{a_{k+1}} a_i^{(k)} f_i$$

and still have $||h_k''|| \leq 2$ and $h_k''(z_j) = \delta_{kj}$. Let $x = \sum_{i=1}^{\infty} t_i e_i$. We write $x = \Sigma x_k$ where $x_k = \sum_{a_k+1}^{a_{k+1}} t_i e_i$. The desired projection is then

$$P(x) = \sum_{k=1}^{\infty} h_k''(x_k) z_k.$$

P is clearly the identity on $\overline{\text{span}}\{z_k\}$, therefore, it suffices to show that P is well defined or, using the fact that c_o is block homogeneous, that the sequence $\{h_k''(x_k)\} \in c_o$. But $x = \sum_{k=1}^{\infty} x_k$ hence, $||x_k|| \to 0$. Using the fact that $||h_k''|| \leq 2$ for all k, the result is obvious.

The proof for the case of ℓ_p, $(1 \leq p < \infty)$, is similar. We may apply the same method arriving at functionals h_k'' and letting $x = \sum_{k=1}^{\infty} x_k$ we again define $P(x) = \Sigma h_k''(x_k) z_k$. However, we must now show $\{h_k''(x_k)\}$ to belong to the appropriate ℓ_p. But

$$\Sigma |h_k''(x_k)|^p \leq \Sigma ||h_k''||^p ||x_k||^p \leq 2^p \Sigma \sum_{a_k+1}^{a_{k+1}} |t_i|^p = 2^p \sum_{i=1}^{\infty} |t_i|^p < \infty. \blacksquare$$

(5.4.23) COROLLARY

The ideal of compact operators is the only nonzero proper closed two sided ideal in $B(\ell_p)$, $1 \leq p < \infty$, and $B(c_o)$.

Proof. Theorem 5.4.20 and Lemmas 5.4.21 and 5.4.22.■

It is interesting, in the light of the above results, to ask whether there are any other Banach spaces besides ℓ_p, $1 \leq p < \infty$ and c_o which have bases which are perfectly homogeneous and satisfy (+). In fact there are not. M. Zippin [70] has shown that perfect homogeneity alone characterizes the spaces ℓ_p, $1 \leq p < \infty$, and c_o.

5.5 Perturbation ideals

The following is due to A. Lebow and M. Schechter [45]. An earlier more restricted version of this idea was given in [69, p. 622].

(5.5.1) DEFINITION

Let S be a subset of a Banach space A. The perturbation class associated with S is denoted $P(S)$ and

$$P(S) = \{a \in A \mid a + s \in S \text{ for all } s \in S\}.$$

In this section we shall assume that S satisfies

(5.5.2) $\qquad \alpha S \subset S$ for each scalar $\alpha \neq 0$.

(5.5.3) LEMMA

$P(S)$ is a linear subspace of A. If S is an open subset of A, then $P(S)$ is closed.

Proof. Suppose $a, b \in P(S)$, $s \in S$ and α is a scalar $\neq 0$. Then $\alpha a + s = \alpha(a + \frac{s}{\alpha}) \in S$ and $(a + b) + s = a + (b + s) \in S$. Thus $P(S)$ is a subspace. Now assume S is open. Then for each $s \in S$ there is a $\delta > 0$ such that $||c - s|| < \delta$ implies that $c \in S$. If $\{x_n\}$ is a sequence of elements of $P(S)$ converging to an element $x \in A$, then for n sufficiently large $||x_n - x|| < \delta$. Thus $s + x - x_n$ is in S for n large. Since $x_n \in P(S)$, $s + x$ is in S. Thus $x \in P(S)$.■

(5.5.4) LEMMA

Let S_1 and S_2 be subsets of A which satisfy (5.5.2). Assume that S_1 is open, that $S_1 \subset S_2$ and that S_2 does not contain any boundary points of S_1. Then $P(S_2) \subset P(S_1)$.

Proof. Suppose $s_1 \in S_1$ and $a_2 \in P(S_2)$. Then

$$\alpha a_2 + s_1 = \alpha(a_2 + \frac{s_1}{\alpha}) \in S_2 \quad \text{for all scalars } \alpha \neq 0.$$

Since S_1 is open, $\alpha a_2 + s_1 \in S_1$ for $|\alpha|$ sufficiently small. It follows that $\alpha a_2 + s_1 \in S_1$ for all α; otherwise for some α_o the element $\alpha_o a_2 + s_1$ would be a boundary point of S_1 which is in S_2. Thus $a_2 + s_1 \in S_1$.■

We now assume that A is a Banach algebra with identity e. Let G denote the group of invertible elements of A.

(5.5.5) LEMMA

If $GS \subset S$, then $P(S)$ is a left ideal. If $SG \subset S$, then $P(S)$ is a right ideal.

Proof. Suppose $a \in G$, $b \in P(S)$ and $s \in S$. Then

$$ab + s = a(b + a^{-1}s) \in S.$$

Consequently $ab \in P(S)$. Now every element of A is the sum of two elements of G, the first statement follows from Lemma 5.5.3. The second statement is proved in a similar way.■

We now have:

(5.5.6) THEOREM

If S is an open subset of A which satisfies $GS \subset S$ and $SG \subset S$ then $P(S)$ is a closed, two sided ideal.

The radical $\mathcal{R}$ of A may be defined [54] as

$$\mathcal{R} = \{f \in A \mid e + af \in G \text{ for all } a \in G\}.$$

We now have the well known result

(5.5.7) THEOREM

$$P(G) = \mathcal{R}.$$

Proof. Suppose $f \in P(G)$ and $a \in G$. Then $a^{-1} + f \in G$. Hence $e + af = a(a^{-1} + f) \in G$. This means that $f \in \mathcal{R}$. Conversely, suppose $f \in \mathcal{R}$. If $a \in G$, then $e + a^{-1}f \in G$. Hence $a + f = a(e + a^{-1}f) \in G$. Thus $f \in P(G)$.■

In particular from the last two theorems we see $\mathcal{R}$ is a closed two sided ideal of A.

Let $G_\ell(G_r)$ denote the set of left (right) invertible elements of A, and let $H_\ell(H_r)$ denote the set of elements of A that are not left (right) topological divisors of zero. We have

(5.5.8) THEOREM

$$P(H_\ell) \subset P(G_\ell) = \mathcal{R}$$

$$P(H_r) \subset P(G_r) = \mathcal{R}.$$

Proof. Since the boundary of G contains only topological divisors of zero (Theorem 2.4.1), we have by Lemma 5.5.4

$$P(H_\ell) \subset \mathcal{R}, \qquad P(G_\ell) \subset \mathcal{R},$$

$$P(H_r) \subset \mathcal{R}, \qquad P(G_r) \subset \mathcal{R}.$$

To complete the proof we need only show that $\mathcal{R} \subset P(G_\ell)$. If $f \in \mathcal{R}$ and $a \in G_\ell$, then $e + bf \in G$, where b is the left inverse of a. Consequently $(e + bf)^{-1}b(a + f) = e$. This shows that $f \in P(G_\ell)$. A similar argument shows that $\mathcal{R} \subset P(G_r)$.■

Let X be a Banach space and $A = C(X) = \frac{B(X)}{K(X)}$ and π be the natural homomorphism of $B(X)$ onto A.

From Chapters 3 and 4

$$\Phi(X) = \pi^{-1}(G), \quad \Phi_\ell(X) = \pi^{-1}(G_\ell) \quad \text{and} \quad \Phi_r(X) = \pi^{-1}(G_\ell).$$

By Theorem 5.5.8 we have

(5.5.9) THEOREM

For any Banach space X,

$$P(\Phi(X)) = P(\Phi_\ell(X)) = P(\Phi_r(X)) = \pi^{-1}(R) = I(X). \quad \text{(See section 2.5).}$$

5.6 Containment of ideals

We have defined many ideals in $B(X)$ and it is natural to ask what kind of containment relationships exist between these ideals. In general, it may not be possible to tell whether $I \subset J$ or vise versa, without possibly putting some restrictions on X. However, a few containments are true for any X and we present them first.

(5.6.1) LEMMA

Let $I(X)$ be the ideal of inessential operators on X (see section 2.5 and Theorem 5.5.9). Any ideal J of $B(X)$ which lies in $R(X)$ is contained in $I(X)$. (Recall that $R(X)$ denotes the set of Riesz operators, see Chapter 3).

Proof. Since $I(X) = \pi^{-1}[\text{Rad } B(X)/K(X)]$ it follows that $I(X) \subset R(X)$. Since $J \subset R(X)$ it follows that $\pi(J)$ is a two sided ideal in $B(X)/K(X)$ consisting of quasi-nilpotent elements. Therefore if $x \in J$ then $\frac{1}{\lambda}\pi(x)$ is quasi-regular for all $\lambda \neq 0$, in particular $\pi(x)$ is quasi-regular. It follows that $\pi(J)$ is a quasi-regular ideal in $B(X)/K(X)$ and therefore $(J) \subset \text{Rad}[B(X)/K(X)]$. Hence

$$J \subset \pi^{-1}(\pi(J)) \subset \pi^{-1}[\text{Rad } B(X)/K(X)] = I(X). \blacksquare$$

Next we wish to show that $S(X) \subset I(X)$ for any Banach space X. Here, of course, $S(X)$ denotes the ideal of strictly singular operators on X. This result is due to S. R. Caradus [8].

(5.6.2) THEOREM

$S(X) \subset I(X)$ for any Banach space X.

Proof. By Lemma 5.6.1 it suffices to show that $S(X) \subset R(X)$. By Theorem 3.2.2, $R(X) = \{T \in B(X) \mid \lambda I - T$ is Fredholm for all $\lambda \neq 0\}$. Therefore we will show that if $T \in S(X)$, then $\lambda I - T$ is Fredholm for all $\lambda \neq 0$. Theorem 4.4.6 says that $\lambda I - T$ has closed range and finite dimensional null space for all $\lambda \neq 0$. If $|\lambda| > ||T||$, then $(\lambda I - T)^{-1}$ exists, i.e. for $|\lambda| > ||T||$, $\lambda I - T$ is Fredholm of index 0.

Suppose $\lambda I - T$ is not Fredholm for all $\lambda \neq 0$. There exists $\lambda_o \neq 0$ on a ray L such that $\lambda_o I - T$ is not Fredholm and for all $|\lambda| > |\lambda_o|$ on L we have $\lambda I - T$ is Fredholm. We have $\alpha(\lambda_o I - T) < \infty$ and $R(\lambda_o I - T)$ is closed and $\lambda_o I - T$ is not Fredholm. Thus $X/R(\lambda_o I - T)$ is infinite dimensional.

Suppose $N(\lambda_o I - T)$ is n dimensional and let $x_1, x_2, \ldots, x_n$ be a basis for $N(\lambda_o I - T)$. There exist $y_1, y_2, \ldots, y_n \in X$ which are linearly independent modulo $R(\lambda_o I - T)$. Choose $x_i^* \in X^*$ such that $x_i^*(x_j) = \delta_{ij}$ for $i, j = 1,2,\ldots,n$. Define

$$V(x) = \sum_{i=1}^{n} x_i^*(x)y_i.$$

Let $Z = \lambda_o I - T + V$. Z is one to one and has closed range. To see that Z is one to one, suppose $Z(x) = 0$. Then $V(x) = -(\lambda_o I - T)(x)$ and since $y_1, y_2, \ldots, y_n$ are linearly independent modulo $R(\lambda_o I - T)$ we must have $x_i^*(x) = 0$ for $i = 1,2,\ldots,n$. Thus $V(x) = 0$ and $x \in N(\lambda_o I - T)$. Say $x = \sum_{i=1}^{n} c_i x_i$. Then for $1 \leq j \leq n$ we have

$$0 = x_j^*(x) = \sum_{i=1}^{n} c_i x_j^*(x_i) = c_j, \quad \text{or} \quad c_j = 0 \quad \text{for} \quad 1 \leq j \leq n$$

and therefore $x = 0$, i.e. Z is one to one.

$R(Z) \neq X$ since $X/R(\lambda_o I - T)$ is infinite dimensional and V is a finite dimensional operator.

Take $|\lambda_1| > |\lambda_o|$ sufficiently close in on ray L so that by Theorem 2.5.6(b) we have $\lambda_1 I - T + V$ is one to one and has closed range $\neq X$. $i(\lambda I - T)$ is a continuous function of λ and since for large λ, $i(\lambda I - T) = 0$ we have $i(\lambda_1 I - T) = 0$. V is compact and therefore by Chapter 2, $\lambda_1 I - T + V$ is Fredholm and $i(\lambda_1 I - T + V) = i(\lambda_1 I - T) = 0$. But $\alpha(\lambda_1 I - T + V) = 0$, therefore $\beta(\lambda_1 I - T + V) = 0$, a contradiction.∎

We now look at the possibility of $S(X) = I(X)$. As we will see this is not always the case, but for particular X the equality holds.

(5.6.3) DEFINITION

A Banach space X is <u>subprojective</u> if, given any closed infinite dimensional subspace M of X, there exists a closed infinite dimensional subspace N contained in M and a continuous projection of X onto N.

Subprojective spaces were investigated by R. J. Whitley [67].

(5.6.4) LEMMA

Let X be any Banach space. Then $T^* \in I(X^*)$ implies that $T \in I(X)$.

Proof. By Theorem 5.5.9, $I(X) = \{T \in B(X) | T + U \in \Phi(X)$ for all $U \in \Phi(X)\}$. Therefore suppose $T \notin I(X)$, then there exists a $U \in \Phi(X)$ such that $T + U \notin \Phi(X)$. By Proposition 1.2.7, $W \in \Phi(X)$ if and only if $W^* \in \Phi(X^*)$, so $T^* + U^* \notin \Phi(X^*)$ and $U \in \Phi(X)$ implies $U^* \in \Phi(X^*)$. Therefore $T^* \notin I(X^*)$. So $T^* \in I(X^*)$ implies $T \in I(X)$.■

The following result is due to W. E. Pfaffenberger [52].

(5.6.5) THEOREM

Let X be any subprojective Banach space. Then $S(X) = I(X)$.

Proof. By Theorem 5.6.2 $S(X) \subset I(X)$.

To prove the converse take $T \in I(X)$ and suppose $T \notin S(X)$. Hence there exists a closed infinite dimensional subspace $X_1 \subset X$ such that T is a homeomorphism of X_1 onto $T(X_1)$. $T(X_1)$ is a closed infinite dimensional subspace of X, so since X is subprojective there exists an infinite dimensional closed complemented subspace $X_2 \subset T(X_1)$, where $X = X_2 \oplus X_3$. (This is equivalent to there being a continuous projection of X onto X_2). Define $\tilde{T} \in B(X)$ by $\tilde{T} \equiv T^{-1}$ on X_2 and $\tilde{T} \equiv 0$ on X_3. Since $T \in I(X)$ we know $\pi(T) \in \text{Rad}[B(X)/K(X)]$, so $I - WT \in \Phi_+(X)$ for all $W \in B(X)$, in particular $I - \tilde{T}T \in \Phi_+(X)$. This implies that $\alpha(I - \tilde{T}T) < \infty$. $\tilde{T}$ is a homeomorphism on X_2, so $\tilde{T}(X_2)$ is an infinite dimensional subspace of X and $\tilde{T}T\tilde{T}(X_2) = \tilde{T}(X_2)$, since $T\tilde{T}$ is the identity on X_2. Therefore $(I - \tilde{T}T)(\tilde{T}(X_2)) = 0$, which implies $\tilde{T}(X_2) \subset N(I - \tilde{T}T)$ which implies $\alpha(I - \tilde{T}T) = \infty$, a contradiction. Therefore $T \in S(X)$, so $I(X) \subset S(X)$ for X subprojective.■

We finish this chapter by investigating the containment relationships between the ideals $K(X)$, $S(X)$, $P(\Phi_+(X))$, $P(\Phi_-(X))$ and $I(X) = P(\Phi(X))$.

(5.6.6) THEOREM

$T \in \Phi_+(X)\backslash\Phi(X)$ implies $T \notin \overline{\Phi(X)}$.

Proof. Suppose $T \in \Phi_+(X)\backslash\Phi(X)$. By Theorem 4.2.1 there exists an $\varepsilon > 0$ such that if $U \in B(X)$ and $||U|| < \varepsilon$ then $T + U \in \Phi_+(X)\backslash\Phi(X)$. Therefore $T \notin \overline{\Phi(X)}$.∎

(5.6.7) COROLLARY

$T \in \Phi_-(X)\backslash\Phi(X)$ implies $T \notin \overline{\Phi(X)}$.

Proof. Use the relationships between T and T^*.∎

We now give a restatement of Theorem 4.4.6.

(5.6.8) THEOREM

$S(X) \subset P(\Phi_+(X))$ for any Banach space X.

Theorems 4.4.4 and 5.6.8 combine with the results of Chapter 4 and section 5.5 to give

(5.6.9) THEOREM

For any Banach space X the following containments always hold

a. $K(X) \subset S(X) \subsetneq P(\Phi_+(X)) \subset P(\Phi(X)) = I(X)$

b. $K(X) \subset P(\Phi_-(X)) \subset P(\Phi(X)) = I(X)$.

(The containments $P(\Phi_-(X)) \subset P(\Phi(X))$ and $P(\Phi_+(X)) \subset P(\Phi(X))$ follow from Theorems 5.5.4 and 5.6.6 and Corollary 5.6.7 and the fact that the index is continuous and integer valued on $\Phi(X)$).

The example below was given in [28] and shows that, in general, no two ideals among those in Theorem 5.6.9 (except possibly $S(X)$ and $P(\Phi_+(X))$ for which the problem remains open) coincide, and that the ideal $S(X)$ is not always contained in $P(\Phi_-(X))$.

(5.6.10) EXAMPLE

Let $X = \ell_q \times L_p$, where $L_p = L_p(-1,1)$ and $1 < p < q < 2$. Since ℓ_q has smaller linear dimension than L_p [37], there exists an operator

V_1 which maps ℓ_q onto a subspace N of $L_p(-1,0)$. Define $V \in B(X)$ by $V(x,y) = (0,V_1(x))$. Choose $A \in B(X)$ such that A coincides with V on ℓ_q and A maps $L_p(-1,1)$ isomorphically onto $L_p(0,1)$. A is a Φ_+ operator but since $A - V$ vanishes on ℓ_q, $A - V$ is not a Φ_+ operator, so $V \notin P(\Phi_+(X))$.

The conjugate space of X is $X^* = \ell_{q'} \times L_{p'}$ where $p' = \frac{p}{p-1}$ and $q' = \frac{q}{q-1}$. The operator V^* is given by

$$V^*(w,z) = (V_1^*(z),0).$$

We will show that V^* is strictly singular. For if $V^* \notin S(X^*)$ there is an infinite dimensional closed subspace $M \subset X^*$ such that V^* is a homeomorphism of M and $V^*(M)$. Let Q be the projection of X^* onto $L_{p'}$ parallel to $\ell_{q'}$. If $z \in Q(M)$, there is a $w \in \ell_{q'}$ such that $(w,z) \in M$. For some $a > 0$,

$$|V_1^*(z)| = |(V_1^*(z),0)| = |V^*(w,z)| \geq a(|w| + |z|) \geq a|z|.$$

Hence V_1^* is a homeomorphism of $Q(M)$, an infinite dimensional closed subspace of $L_{p'}$, and a subspace of $\ell_{q'}$. Every infinite dimensional closed subspace of $\ell_{q'}$ contains a subspace isomorphic to $\ell_{q'}$, [2]. Hence V_1^* determines a homeomorphism of a subspace of $L_{p'}$ and $\ell_{q'}$. Since $2 < q' < p'$, this is impossible [51].

Since X is reflexive, it follows that $V \in P(\Phi_-(X))$ and $V^* \notin P(\Phi_-(X^*))$.

Therefore, $P(\Phi(X)) \neq P(\Phi_+(X))$, $P(\Phi_-(X)) \neq K(X)$, $P(\Phi_+(X^*)) \neq P(\Phi_-(X^*))$, $S(X^*) \neq K(X^*)$ and $P(\Phi(X^*)) \neq P(\Phi_-(X^*))$.

Moreover, $S(X^*)$ is not contained in $P(\Phi_-(X^*))$ and although $V^* \in S(X^*)$, its adjoint V is not in $S(X)$.

This example also shows that Theorem 5.6.5 cannot be generalized to reflexive spaces (in fact X is also superprojective [67]), since $V^* \in S(X^*) \subset I(X^*)$ we have that $V \in I(X)$ by Lemma 5.6.4, but $V \notin S(X)$, so this implies $S(X) \neq I(X)$. We note that this example also shows that a converse to [67, Theorem 2.2, p. 254] is not possible, since $X^* = \ell_{q'} \times L_{p'}$, with $2 < q' < p'$ implies X^* is subprojective, but $V^* \in S(X^*)$ and $V \equiv V^{**} \notin S(X^{**})$.

Chapter 6

GENERALIZATIONS OF FREDHOLM THEORY

Previously we have considered the Banach algebra $B(X)$ and the ideal $K(X)$ of compact operators with the major aim of exploiting the properties of the canonical homomorphism $B(X) \to B(X)/K(X)$. We will now consider other algebras in which a suitable ideal can be found; in fact, to begin with, we will look at a development where we have only a ring rather than an algebra. This work, due to B. A. Barnes [3], considers rings in which the "socle" is defined. The socle then provides an appropriate two sided ideal for Fredholm theory.

The second section of this chapter considers studies by B. Gramsch [30] and L. A. Coburn and A. Lebow [12] on Fredholm theory of a very general kind, the ideal being chosen arbitrarily. In order to obtain substantial new results, however, it is natural to consider operator algebras. In particular, the work of M. Breuer ([5], [6]) on Fredholm theory in von Neumann algebras is expounded in some detail. As applications of these concepts, we consider the contributions of R. G. Douglas and various other workers ([9], [10], [11], [19], [20]) to the theory of Toeplitz operators and the Wiener-Hopf equation. Finally, we give an exposition of recent work of D. G. Schaeffer concerning the application of Breuer's theory to finite difference equations.

The first five chapters of this book have been reasonably self contained but it will now be necessary, in order to keep the exposition to an appropriate length, to omit some proofs entirely and in some other cases, to give merely a sketch of a proof. In all such situations, we have taken care to direct the reader to sources where further information can be obtained.

6.1 Fredholm theory in semiprime rings

We begin with some elementary ideas from ring theory. Let A denote a ring with identity 1 and suppose that I is an ideal in A. (We shall use the word "ideal" to denote either a left or right ideal; as usual, some definitions and statements of theorems will be valid for either left ideals or right ideals. We will sometimes state only the left ideal version and leave the reader to observe that the other version is also possible.) We will write I^n to denote the set of products $x_1 \cdot x_2 \cdots x_n$ where each x_i belongs to I. An ideal is called <u>nilpotent</u> if $I^n = \{0\}$ for some integer n. A ring which contains no nilpotent ideals is called <u>semiprime</u>.

For any subset B of a ring A we write $L(B)$ to denote the set

$$\{x \in A : xb = 0 \quad \text{for all} \quad b \in B\}.$$

Evidently $L(B)$ is a left ideal in A and, if A is semiprime, $L(A)$ must be the zero ideal since, by definition, $L(A)^2 = \{0\}$. Similarly we define $R(B)$ and make the same observation about $R(A)$.

A semiprime ring has the important property that its minimal ideals can be conveniently described.

(6.1.1) LEMMA

Let A be a semiprime ring containing ideal I. Then

(i) I is a minimal left (right) ideal in A if and only if there exists an idempotent e in I such that $I = Ae$ ($I = eA$) and eAe is a division ring.

(ii) Ae is a minimal left ideal if and only if eA is a minimal right ideal.

Proof.

(i) Suppose I is a minimal left ideal. Since $I^2 \neq \{0\}$, there exists b in I such that $Ib \neq \{0\}$. But Ib is a left ideal and $Ib \subset I$. By assumption, $Ib = I$. Hence there exists e in I such that $eb = b$. Then $e^2b = eb$ so that $(e^2 - e)b = 0$. Now $L(b)$ is a left ideal and $I \cap L(b)$ is a left ideal contained in I. This containment must be proper for otherwise $L(b)$ would contain I so that Ib would equal $\{0\}$; but this is false. Hence by the minimality of I, $I \cap L(b) = \{0\}$. Now $e^2 - e \in I \cap L(b)$ so that $e = e^2$. Also $e \neq 0$ so that Ae is a nonzero left ideal and $Ae \subset I$. Therefore $Ae = I$.

Now we show that eAe is a division ring. Let eae be any nonzero element of eAe. Then Aeae = Ae because of minimality so that (eAe)(eae) = eAe and hence there exists ebe in eAe such that (ebe)(eae) = e, i.e. eae has a left inverse in eAe. Therefore eAe is a division ring.

Conversely, let e be an idempotent such that eAe is a division ring. Now since eAe is a division ring, there exists a in A such that $ae \neq 0$. Now $R(A) = \{0\}$ so that $Aae \neq \{0\}$ and hence $(Aae)^2 \neq \{0\}$ since Aae is a left ideal. Therefore there exists b in A such that $ebae \neq 0$. Since eAe is a division ring, there exists ece such that $(ece)(ebae) = e$. Thus, for every d in A, $de = decebae$. Thus $Ae = Aae$ for every a in A such that $ae \neq 0$. Now let I be a left ideal such that $\{0\} \neq I \subset Ae$. Then choose a such that $ae \in I$. Then $Ae = Aae \subset AI \subset I$. Hence $Ae = I$ so that Ae is minimal.

The proof of (i) for right ideals is exactly analogous. Finally, it is clear that (ii) is a consequence of (i).■

REMARK

The idempotents which give the minimal ideals are called minimal idempotents. The sum of the minimal left (right) ideals is called the left (right) socle, denoted by $S_\ell (S_r)$. We now show that, in a semiprime ring, the left and right socles coincide.

(6.1.2) LEMMA

If A is a semiprime ring, then $S_\ell = S_r$ and the common value S, the socle, is a two sided ideal.

Proof.

We will show that S_ℓ is a two sided ideal. Obviously S_ℓ is a left ideal. Now consider a minimal left ideal $I = Ae$. For each a in A, Ia is a left ideal. If $Ia \neq \{0\}$, choose $b \in I$ such that $ba \neq 0$. Now by the minimality of I, I must be the smallest left ideal containing b. It is easy to deduce from this that Ia is the smallest left ideal containing ba. Hence Ia is a minimal left ideal and therefore $Ia \subset S_\ell$. Hence $S_\ell a \subset S_\ell$ for all $a \in A$ so S_ℓ is a right ideal. Thus we see that S_ℓ is a two sided ideal. Similarly for S_r. Finally, since $e \in S_\ell$ for each minimal idempotent e, we have $eA \subset S_\ell$ and hence $S_r \subset S_\ell$. Similarly $S_\ell \subset S_r$, so the proof is complete.■

EXAMPLE

Let X be a Banach space and let $A = B(X)$. It is clear that A is semiprime since every ideal contains finite dimensional operators (Theorem 5.2.1); moreover the minimal idempotents are just the one dimensional projections. Thus the minimal ideals consist of one dimensional operators and the socle is $F(X)$, the ideal of finite dimensional operators. Since the Fredholm operators can be characterized as the inverse image of the invertible elements in $B(X)/F(X)$ (see the proof of Lemma 3.2.6), we have evidence that the socle is an appropriate choice in the ring-theoretic setting.

We now consider the concept of order for an ideal, a notion which plays the role of dimension in ring theory.

DEFINITION

A left ideal I has order n if I can be written as the sum of n minimal left ideals but of no fewer. For convenience, we define the order of the zero ideal to be zero and write $\theta(I)$ for the order of an ideal I.

As one might expect, it is possible to show that if I is a non zero left ideal of order n, then every maximal set of mutually orthogonal minimal idempotents in I contains n elements and if $\{e_1, e_2, \ldots, e_n\}$ is such a set then $I = Ae$ where $e = e_1 + e_2 + \cdots + e_n$.

DEFINITION

Let A be a semiprime ring with 1 and with socle S. An element x of A is called Fredholm if, for some y_1 and y_2 in A,

$$y_1x - 1 \in S$$

and

$$xy_2 - 1 \in S.$$

(We assume throughout that all rings have unit elements.)

(6.1.3) THEOREM

If A is a semiprime ring, then x is a Fredholm element if and only if $Ax = A(1 - e)$ and $xA = (1 - f)A$ for some idempotents e and f in S.

Proof. We will do half the proof, leaving the other because it is similar. Suppose $Ax = A(1 - e)$ for some idempotent e in S. Then it follows that $A = Ax + Ae$ so that there exist $v, w \in A$ such that $vx + we = 1$. Hence $vx - 1 = -we \in S$. Conversely, assume that x is such that $vx - 1 = s$ with $v \in A$ and $s \in S$. Then $Avx = A(1 + s)$ so that $A(1 + s) \subset Ax$.

Hence $R(Ax) \subset R(A(1 + s))$. Now suppose $u \in R(A(1 + s))$ so that $a(1 + s)u = 0$ for all a. But in a semiprime ring $R(A) = \{0\}$ so that $(1 + s)u = 0$, i.e. $u = -su$. Hence $R(A(1 + s)) \subset sA$. Now sA is a right ideal in S and all such ideals are of the form eA for some idempotent e. It is now possible to show that this idempotent e has the required property: $Ax = A(1 - e)$. The proof, however, is somewhat technical and the reader is referred to [3] for details. (It is interesting to note in passing that this theorem is by no means obvious even in the familiar setting $A = B(X)$).■

The Index. Suppose u is a Fredholm element in semiprime ring A. Then $L(uA)$ and $R(Au)$ are ideals in S as the above proof has made clear. Hence they both have finite order since each ideal in S is the sum of a finite number of minimal ideals. We therefore define the index $\kappa(u)$ as follows

$$\kappa(u) = \theta[(uA)] - \theta[(Au)].$$

It is then possible to prove an index theorem:

(6.1.4) THEOREM

If A is a semiprime ring and u,v are Fredholm elements in A then so is uv; moreover $\kappa(uv) = \kappa(u) + \kappa(v)$.

The proof of this theorem is a rather long technical exercise in manipulating idempotents and ideals. Once again, the reader is referred to [3]. Finally we note that when A is a semiprime Banach algebra, then the index is a continuous function on the open semigroup of Fredholm elements [3, Theorem 4.1].

6.2 Generalized index theory

If A is any ring with unit I and $\mathcal{I}$ is a two sided ideal, we will write $\pi : A \to A/\mathcal{I}$ to denote the canonical homomorphism, G to denote the group of invertibles in $A/\mathcal{I}$ and F to denote the semigroup $\pi^{-1}(G)$.

An (abstract) index consists of a homomorphism i of the semigroup F into the additive group Z of integers such that

(a) $i(T) = 0$ for all invertible elements T in A

(b) $i(I + K) = 0$ for all K in $\mathcal{I}$.

From the above definition, it is easy to deduce that $i(T + K) = i(T)$ for each $T \in F$, $K \in I$. For if $T \in F$, then there exists $S \in F$ and $K_1 \in I$ such that $ST = I + K_1$. Hence $S(T + K) = I + K_1 + SK$ so that $i(S) + i(T + K) = i(I + K_1 + SK) = i(I) = 0$. But $i(ST) = i(S) + i(T) =$ $= i(I + K_1) = i(I) = 0$ so by comparing these sets of equalities, $i(T + K) = i(T)$.

In the above setting, two special kinds of ideals were studied by B. Gramsch [30].

(6.2.1) DEFINITION

An ideal J is called an <u>R-ideal</u> if $I + J \subset F$. A two sided ideal J is called an R^+-ideal if $\hat{\pi}^{-1}(\hat{G}) = F$ where $\hat{\pi}$ is the canonical homomorphism $A \to A/J$ and $\hat{G}$ is the group of invertibles in A/J.

The above notions are related to the Riesz operators and to the ideal of inessential operators.

(6.2.2) LEMMA

If J is a two sided R-ideal which contains I, then J is an R^+-ideal.

Proof. Because $J \supset I$, it is easy to see that $F \subset \hat{\pi}^{-1}(\hat{G})$. To prove the reverse inclusion, consider any element T of $\hat{\pi}^{-1}(\hat{G})$. Then there exist $L, R \in \hat{\pi}^{-1}(\hat{G})$ with $LT = I + K_1$ and $TR = I + K_2$ for some K_1 and K_2 in J. But since J is an R-ideal, $I + K_1$ and $I + K_2$ lie in F. Hence there exist L_1 and R_1 in F such that $L_1(I + K_1) = I + K_3$ and $(I + K_2)R_1 = I + K_4$ with K_3 and K_4 in I. Thus $L_1LT = I + K_3$ and $TRR_1 = I + K_4$ so that $\pi(T)$ is invertible in A/I, i.e. $T \in F$. This completes the proof.∎

(6.2.3) THEOREM

Let A and I be defined as before. Then there exists a unique maximal R-ideal Q in A. Moreover, Q is an R^+-ideal and A/Q is semisimple (i.e. has zero radical). In fact, if J is any R^+-ideal, then

$$Q = \hat{\pi}^{-1} \operatorname{Rad}(A/J)$$

and hence

$$Q = \pi^{-1} \operatorname{Rad}(A/I).$$

Proof. Suppose we define $\mathcal{Q} = \pi^{-1}\, \mathrm{Rad}(A/\mathcal{I})$. Then it is easy to verify that $\mathcal{Q}$ is a two sided R-ideal containing $\mathcal{I}$. By the lemma, $\mathcal{Q}$ is therefore an R^{+}-ideal. Suppose now that $\mathcal{K}$ is an arbitrary left R-ideal. Then $I + K \in \mathcal{F}$ for all K in $\mathcal{K}$ so that $\pi(I + K)$ is invertible. Then, because $\mathcal{K}$ is an ideal, $\pi(I + TK)$ is invertible for each T in A, K in $\mathcal{K}$. Thus $\pi(K) \in \mathrm{Rad}(A/\mathcal{I})$ so that $\mathcal{K} \subset \mathcal{Q}$; i.e. $\mathcal{Q}$ is a maximal R-ideal. For the proof of the remaining assertion, let $\mathcal{J}$ be an R^{+}-ideal and write $\hat{\mathcal{Q}} = \hat{\pi}^{-1}[\mathrm{Rad}(A/\mathcal{J})]$. Then $\hat{\pi}(I + TS)$ is invertible for all $T \in \hat{\mathcal{Q}}$, $S \in A$. Hence, by definition of R^{+}-ideals, $\pi(I + TS)$ is invertible for all $T \in \hat{\mathcal{Q}}$, $S \in A$ so that $\hat{\mathcal{Q}} \subset \mathcal{Q}$. Conversely, if $T \in \mathcal{Q}$, then $TS \in \mathcal{Q}$ for all $S \in A$ and therefore $I + TS \in \mathcal{F} = \hat{\pi}^{-1}(\hat{\mathcal{G}})$. Thus $\hat{\pi}(I + TS) \in \hat{\mathcal{G}}$ so that $\hat{\pi}(T) \in \mathrm{Rad}(A/\mathcal{J})$, i.e. $T \in \hat{\mathcal{Q}}$.■

REMARKS

(1) In the case where X is a Banach space and $A = B(X)$ and $\mathcal{I} = K(X)$, the ideal $\mathcal{Q}$ is exactly the ideal of inessential operators.

(2) In the general case, $\mathcal{Q}$ can be identified in another way. For this we need the concept of a primitive ideal: a two-sided ideal $\mathcal{P}$ is called primitive if there exists a linear space Y, a homomorphism ϕ of A into $\mathcal{L}(Y)$ with $\mathcal{P}$ as its kernel such that $\{0\}$ and Y are the only invariant subspaces for the family of operators $\phi(A)$, i.e. primitive ideals are the kernels of strictly irreducible representations. For any subset $\tilde{A}$ of A, the hull $h(\tilde{A})$ of $\tilde{A}$ is defined to be the set of primitive ideals which contain $\tilde{A}$; for any set ω of primitive ideals, the kernel $k(\omega)$ of ω is defined to be the intersection of all the ideals in ω. The radical can be shown to equal $k(\Omega)$ where Ω is the set of all primitive ideals in A. Once this fact is established, it is a simple exercise to show that

$$\mathcal{Q} = k(h(\mathcal{I})).$$

Next, we consider the case where A is a Banach algebra and $\mathcal{I}$ is a closed two-sided ideal in A. Write Z to denote the additive group of integers with the discrete topology. It is possible to show that every index is continuous. In fact, more is true:

(6.2.4) THEOREM

Every semigroup homomorphism of $\mathcal{F}$ into Z is continuous.

Proof. Let κ be such a homomorphism. Because of the continuity of $\pi : A \to A/I$, it will suffice to show that any group homomorphism h of G, the group of invertibles in A/I, into Z is continuous. Moreover, because of the nature of Z, it will suffice to show that h is identically zero on some neighbourhood of the identity 1. Let $||x - 1|| < 1$; then $\log x$ is defined by $\sum_{1}^{\infty} (-1/K)^{K-1}(x - e)^K$ and for every x, $\exp x$ is defined by the usual series. Not surprisingly, $\exp(\log x) = x$ and $\exp(a + b) = \exp a \exp b$ when a and b commute. Consider $x_n = \exp(1/n \log x)$. Then $x = x_n^n$ and $h(x) = n\, h(x_n)$. But if $h(x) \neq 0$ then $h(x)$ is an integer which is divisible by n for each n. Clearly, we must conclude that $h(x) = 0$ for $||x - 1|| < 1$.■

Various other results along the same lines can be found in Gramsch [30]. Another approach is given by Coburn and Lebow [12] who take the following very general viewpoint: let S be an open semigroup of a topological algebra A. An <u>index</u> is then defined as a homomorphism α from S to any semigroup $\mathcal{D}$ such that α is constant on the connected components of S. Every such semigroup S gives rise to another semigroup $H(S)$ whose elements are the components of S and whose multiplication is defined as follows: if C_1 and C_2 are components of S and $x_1 \in C_1$ and $x_2 \in C_2$, then define the product $C_1 \circ C_2$ to be the component of S to which $x_1 x_2$ belongs. It is a simple exercise to show that this definition is independent of the choice of x_1 and x_2. From each index α, we obtain a semigroup homomorphism $\alpha_* : H(S) \to \mathcal{D}$ induced in the obvious way. From such ideas, one can investigate the relationship between the number of components on which an index has a given value and the connectivity of the group of invertibles in A. Consider a Banach algebra A with identity 1 and let I be a closed two sided ideal. Suppose that F is the set of Fredholm elements given by A and I. We will write $G(A)$ to denote the group of invertibles in A. Then, from Chapter 2 we have $G_0(A)$, the component of $G(A)$ which contains the identity, is a normal subgroup, the only open connected subgroup of $G(A)$ and each other component of $G(A)$ can be written as $xG_0(A)$ for some x in $G(A)$. We will consider the action of π, the canonical homomorphism $A \to A/I$, on $H(F)$ and $H(G(A))$.

(6.2.5) LEMMA

(a) The mapping π gives rise to an isomorphism ϕ of $H(F)$ onto $H(G(A/I))$.

(b) If the elements of I have connected resolvent set, then the mapping π gives rise to an isomorphism ψ of $H(G(A))$ into $H(G(A/I))$.

Proof.

(a) Let C be a component of F. Then since π is an open continuous map, $\pi(C)$ is an open connected set. Therefore, there must be a component $\mathcal{C}$ of $G(A/I)$ such that $\mathcal{C} \supset \pi(C)$. Suppose that this inclusion is proper and let $x \in \mathcal{C}\backslash\pi(C)$. Then $x = \pi y$ for some $y \in F$ and y must belong to some other component C_1 of F. Now $\pi(C)$ and $\pi(C_1)$ must be disjoint; for if $\pi(v) = \pi(v_1)$ for $v \in C$ and $v_1 \in C_1$, then $v_1 - v \in I$ and $tv_1 + (1 - t)v = v + t(v_1 - v) \in F$ for $0 \leq t \leq 1$. But this would mean that v_1 and v belong to the same component. Now consider all components C_K of F such that $\pi(C_K) \subset \mathcal{C}$. We have an open covering of $\mathcal{C}$ consisting of disjoint open sets. Since this is impossible, it must be the case that $\pi(C) = \mathcal{C}$. The map π therefore takes $H(F)$ onto $H(G(A/I))$ and the above argument makes it clear that the mapping is one-to-one. Finally, to see that we have an isomorphism, consider C_1 and C_2 in $H(F)$. Then it is easy to verify that π maps $C_1 \circ C_2$ onto $\pi(C_1) \circ \pi(C_2)$.

(b) Consider the action of π on a component $xG_0(A)$ of $G(A)$. Clearly $\pi[xG_0(A)] = \pi(x)\,\pi[G_0(A)]$ and since π is an open continuous homomorphism, then $\pi[G_0(A)]$ is an open connected subgroup. Therefore $\pi[G_0(A)] = G_0(A/I)$ so that π maps $H(G(A))$ into $H(G(A/I))$. Moreover this mapping is one-to-one. For suppose π maps two components C_1 and C_2 of $G(A)$ onto the same component of $G(A/I)$. Then there is x_1 in C_1 and x_2 in C_2 such that $\pi x_1 = \pi x_2$. Thus $x_1 - x_2 \in I$ so that $1 - x_1^{-1}x_2 \in I$. Now because $1 - x_1^{-1}x_2$ has connected resolvent set, we can find a continuous function ϕ on $[0,1]$ such that $\phi(0) = 0$, $\phi(1) = 1$ and $\frac{1}{\phi(t)} \in \text{res}(1 - x_1^{-1}x_2)$ for all $t > 0$. Then

$1 - \phi(t)(1 - x_1^{-1}x_2) = \phi(t)[\frac{1}{\phi(t)} - (1 - x_1^{-1}x_2)]$ is an arc in $G(A)$ joining $x_1^{-1}x_2$ to 1. Hence $x_1^{-1}x_2 \in G_0(A)$ so that $x_2 \in x_1G_0(A)$, i.e. x_1 and x_2 belong to the same component of $G(A)$. Thus the mapping is an isomorphism.■

(6.2.6) THEOREM

Let A be a Banach algebra with identity containing a closed two sided ideal I whose elements have connected resolvent set. Let F denote the Fredholm elements of A relative to I and suppose that $\mathcal{D}$ is a semigroup with identity e. Let κ denote an index $F \to \mathcal{D}$ such that $\kappa(x) = e$ when x is invertible in A. Then $H(G(A))$ is isomorphic to a subgroup of $\ker \kappa_*$.

Proof. Consider the mappings ϕ and ψ of Lemma 6.2.5. Then $\phi^{-1} \circ \psi$ is an isomorphism of $H(G(A))$ into $H(F)$ which maps each component of $G(A)$ into the component of F which contains it. Since κ is constant on each component of F and $\kappa(x) = e$ for $x \in G(A)$, it is clear that $\phi^{-1} \circ \psi$ has range in $\ker \kappa_*$. This completes the proof.■

COROLLARY

If X is a Banach space and $A = B(X)$ with $I = K(X)$, then $\ker \kappa_*$ is isomorphic to $H(G(A))$ so that κ_* is an isomorphism if and only if $G(A)$ is connected.

Proof. Suppose $\kappa(T) = 0$ for $T \in B(X)$. Then we can write $T = T_1 + C$ with $T_1 \in G(A)$ and $C \in K(X)$. Hence $\ker \kappa = G(A) + K(X)$ so that each component of $\ker \kappa$ must contain a component of $G(A)$. Thus the isomorphism of Theorem 6.2.6 must be onto.■

6.3 Operator algebras

In preparation for the next section of this chapter, we give some definitions and properties of C* algebras of operators on a Hilbert space H. For further information on this subject, the reader is referred to the well known monograph of J. Dixmier, [18].

(6.3.1) DEFINITION

A subalgebra A of B(H) is called *self adjoint* if $T^* \in A$ whenever $T \in A$. A uniformly closed self adjoint subalgebra A of B(H) is called a *C* algebra*.

A commutative C* algebra is an especially tractable object; a fundamental theorem of Gelfand and Naimark states that every such algebra is isometrically *-isomorphic to $C(\Omega)$, the space of continuous functions which "vanish at infinity" on the locally compact Hausdorff space Ω. In the situations to appear later, we will have noncommutative C* algebras and we will factor out the commutator ideal I, i.e. we define I to be the smallest closed two sided ideal containing the set of commutators

$\{TS - ST : S, T \in A\}$ and we consider A/I. It is easy to see that A/I is a commutative algebra. But we can even show that A/I "is" a C* algebra. The following theorem, proved by Calkin for $B(H)/K(H)$, is recorded without proof since, in fact, we will not need to make explicit use of it.

(6.3.2) THEOREM

If A is a C* algebra and I is a closed two sided ideal in A, then $I = I^*$ and A/I is isometrically *-isomorphic to a C* algebra.

Proof. [18, p. 17].■

We now give the proofs of two results which have obvious relevance to Fredholm theory in C* algebras. For the first of these, we need some additional concepts. If A is a C* algebra in B(H), a closed subspace H_0 of H is called <u>reducing</u> for A if every operator T in A satisfies the conditions $TH_0 \subset H_0$ and $TH_0^{\perp} \subset H_0^{\perp}$. An algebra A is called <u>irreducible</u> if H and {0} are the only reducing subspaces.

(6.3.3) THEOREM

Let A be an irreducible C* algebra in B(H). Then if A contains one nonzero compact operator, it contains them all.

Proof. Suppose T is a nonzero compact operator in A. Then $T + T^*$ and $i(T - T^*)$ cannot both be zero. Hence A contains a nonzero self adjoint compact operator K. Now K cannot be quasinilpotent (for quasinilpotent self adjoint operators are zero) so there exists a nonzero eigenvalue λ for K. Let P_λ denote the corresponding spectral projection. Then from the spectral theorem [56, p. 275] we know that P_λ can be expressed as the uniform limit of a sequence of polynomials in K. Hence P_λ is a finite dimensional projection in A. Let E denote a nonzero finite dimensional projection in A with minimum rank and consider the closed subalgebra EAE as a C* algebra on the Hilbert space E(H). Now E(H) is finite dimensional and because E has minimal rank, no element of EAE can have a disconnected spectrum; otherwise we would obtain projections in EAE with rank smaller than the rank of E. Hence EAE must consist of scalar multiples of E. We now show that E must be one dimensional. Suppose in fact that x and y were independent vectors in E(H). Now $\{Tx : T \in A\}$ has closure H_0 which is a reducing subspace for A; obviously $SH_0 \subset H_0$ for all $S \in A$ and $SH_0^{\perp} \subset H_0^{\perp}$ since if $h \in H_0$ and $k \in H_0$ then

$(Sk,h) = (k,S^*h) = 0$ since $S^*h \in H_0$. Thus H_0 must equal H. Hence for some sequence $\{T_n\}$ in A, we must have $T_n x \to y$. Hence $ET_nEx - y = ET_nx - Ey = E(T_nx - y) \to 0$. But we have proved that ET_nE must be a scalar multiple of E, therefore $ET_nE = \lambda_n E$ for some $\lambda_n \in \mathbb{C}$. But then we have $\lambda_n Ex \to y$ which implies $\lambda_n x \to y$. But x and y are independent so we have a contradiction. Thus E must have rank one.

We now show that every rank one operator is in A; clearly the assertion of the theorem will follow. For x and y in H, define $T_{y,x}$ to be the rank one operator given by

$$T_{y,x}(z) = (z,x)y.$$

For a unit vector x_0 in EH, let $T_nx_0 \to y$ for some sequence $\{T_n\}$ in A. Then $\{T_nT_{x_0,x_0}\}$ is a sequence in A and $T_nT_{x_0,x_0} \to T_{y,x_0}$. So $T_{y,x_0} \in A$. Similarly, using adjoints we can prove that $T_{x_0,x} \in A$ and since $T_{y,x} = T_{y,x_0}T_{x_0,x}$, we conclude that $T_{y,x} \in A$. Now every rank one operator must be a multiple of some $T_{y,x}$, so the proof is complete.■

Finally, we have to consider the following: suppose A is a C^* algebra, I is a closed two sided ideal and π is the homomorphism $A \to A/I$. Then we can define the Fredholm elements of A in the manner which is familiar. But we also have the possibility that some of the operators of A are Fredholm operators. It is a pleasant fact that for C^* algebras, the two concepts coincide in the following way.

(6.3.4) THEOREM

Let A be a C^* algebra, I a closed two sided ideal. Then the Fredholm operators of A are those which are invertible in A modulo the ideal $I \cap K(H)$, i.e. if ρ denotes the canonical homomorphism $A \to A/[I \cap K(H)]$ then T is a Fredholm operator in A if and only if $\rho(T)$ is invertible in $A/[I \cap K(H)]$.

Proof. Suppose $\rho(T)$ is invertible in $A/[I \cap K(H)]$; then the usual argument shows T to be Fredholm: there exists S_1 and S_2 in A and K_1 and K_2 in $I \cap K(H)$ such that $TS_1 = I + K_1$ and $S_2T = I + K_2$. But this clearly implies T is Fredholm. Now let us assume that T is a Fredholm operator in A. Then we know that T^* is also Fredholm, as is T^*T. Now the range of T^*T is a closed subspace which reduces the self-adjoint operator T^*T. We therefore have a decomposition

$$H = R(T^*T) \oplus N(T^*T)$$

and from this it follows that T^*T has finite ascent and descent so that, by Lemma 3.4.2, $\lambda = 0$ is an isolated point in the spectrum of T^*T or $\lambda = 0$ belongs to the resolvent set of T^*T. In either case, we can associate with $\lambda = 0$ a projection E with $N(T^*T)$ as its range and by Lemma 3.4.2 again, E is given by the operational calculus. By a wellknown property of C^* algebras [54, p. 185], if λ is in the resolvent set of T^*T, then $(\lambda I - T^*T)^{-1}$ belongs to A. Since E is given by integrating this resolvent operator around a suitable curve, we deduce that $E \in A$. Moreover, E has finite rank so that $E \in A \cap K(H)$.

Now consider $S = T^*T + E$. Then $S \in A$ and it is easy to see that S has an inverse in $B(H)$. Again, therefore, S^{-1} belongs to A and

$$I = (S^{-1}T^*)T + S^{-1}E.$$

But since $S^{-1}E \in A \cap K(H)$, the above equation implies that $\rho(T)$ has a left inverse in $A/[I \cap K(H)]$. Similarly, using the same argument with TT^*, we conclude that $\rho(T)$ has a right inverse. This concludes the proof.■

6.4 C* algebras generated by Toeplitz operators

The theory and applications of the Wiener-Hopf equation

$$y(t) = \lambda x(t) + \int_0^\infty k(t - \tau)x(\tau)d\tau \tag{6.4.1}$$

where x and y are functions defined on $[0,\infty)$, has given rise to a huge literature. (The long article [42] by M. G. Krein and its bibliography give a comprehensive account of the method of solution involving a certain factorization of the kernel function k. A very readable introduction to the origin of the equation in problems of prediction theory for stochastic processes and in diffusion and diffraction problems in semi-infinite media is found in the monograph [50] of B. Noble.)

A more recent approach to the Wiener-Hopf equation has involved methods related to the theory developed in earlier sections of this chapter. The contributions of R. G. Douglas and his collaborators have been highly significant in this regard with the book [19] "Banach Algebra Techniques in Operator Theory" representing a comprehensive treatment of the state of development up to about 1970.

The first step involves the application of the Fourier transform $\mathcal{F}$ to equation (6.4.1). For this purpose, we will assume that k belongs to $L_1(R)$ and that x and y belong to $L_2(R^+)$ where R^+ denotes the half line $[0, \infty)$. The Paley-Wiener theorem [34, p. 131] tells us that $\mathcal{F}L_2(R^+)$ consists exactly of the closed subspace $H_2(R)$ of functions in $L_2(R)$ which have an analytic extension into the upper half plane, vanishing at infinity. If we let P denote the orthogonal projection of $L_2(R)$ onto $H_2(R)$, then the application of $\mathcal{F}$ to equation (6.4.1) gives

$$P[(\lambda + \hat{k})\hat{x}] = \hat{y}.$$

The operator $\hat{f} \to P(\hat{k}\hat{f})$ in $H_2(R)$ is called a Toeplitz operator with symbol $\hat{k}$. More generally, if ϕ is any function in $L_\infty(R)$, the corresponding Toeplitz operator W_ϕ in $H_2(R)$ is defined by

$$W_\phi f = P(\phi f).$$

Analogously, one can develop a discrete version of the Wiener-Hopf equation for functions in $\ell_2(Z^+)$ with the convolution given by

$$\sum_{m=0}^{\infty} k_{n-m} f_m.$$

Relative to the standard orthonormal basis, the mapping $\{f_n\} \to (\sum_{m=0}^{\infty} k_{n-m} f_m)$ is represented by an infinite matrix with each diagonal having constant entries, a so-called Toeplitz matrix. The appropriate Fourier transform in this setting merely maps sequences $\{g_n\}$ which are elements of $\ell_2(Z)$ onto the functions g on the unit circle T such that g has $\{g_n\}$ for its sequence of Fourier coefficients. The subspace $\ell_2(Z^+)$ is therefore mapped onto the subspace $H_2(T)$ of $L_2(T)$ consisting of functions whose negative Fourier coefficients are all zero. We again get Toeplitz operators $W_\phi f = P(\phi f)$ where $f \in H_2(T)$, $\phi \in L_\infty(T)$ and P is the orthogonal projection of $L_2(T)$ onto $H_2(T)$. This latter case of Toeplitz operators on the unit circle has been studied by many authors but, in fact, the two cases are equivalent; Devinatz [16] showed that the conformal map of the upper half plane onto the interior of T sets up a unitary equivalence between the two types of Toeplitz operators mentioned thus far.

A more general viewpoint was studied by Douglas and Coburn [9] in the context of abstract harmonic analysis: let G be a locally compact abelian group with dual group $\hat{G}$; fix a sub-semigroup Σ of $\hat{G}$ and let $H_2(\Sigma)$ denote the subspace of $L_2(G)$ consisting of functions f whose Fourier transform $\hat{f}$ has support on Σ. Then Toeplitz operators W_ϕ on $H_2(\Sigma)$ can be defined exactly as before for $\phi \in L_\infty(G)$. In section 6.6, we will pursue this general development in some detail.

The basic problem to be studied concerns the invertibility of Toeplitz operators or more generally the problem of describing the spectrum. To be specific, let us consider Toeplitz operators W_ϕ on the Hilbert space $H = H_2(T)$ with continuous symbol ϕ. Let A denote the C* algebra generated in $B(H)$ by all such operators. Now the map $\phi \to W_\phi$ is clearly linear and $||W_\phi|| \leq ||\phi||$ (in fact, equality holds), but ϕ is not multiplicative so the algebra A is noncommutative. However, we can obtain an important result about its multiplicative structure.

(6.4.2) LEMMA

If ϕ and ψ are in $C(T)$, then $W_\phi W_\psi - W_{\phi\psi}$ is compact.

Proof. Let e_n denote the function $e^{i\theta} \to e^{in\theta}$ in $C(T)$ for each $n \in Z$. Now for $n \geq 0$, we can show that $W_\phi W_{e_n} = W_{\phi e_n}$; for if $f \in H$, then

$e_n f \in H$ and so $W_\phi W_{e_n}(f) = W_\phi(e_n f) = P(\phi e_n f) = W_{\phi e_n}(f)$.

Suppose we now consider $n = -1$. Then

$$\begin{aligned} W_\phi W_{e_{-1}}(f) &= W_\phi P(e_{-1}f) = W_\phi(e_{-1}f - (f,e_0)e_{-1}) \\ &= P(\phi e_{-1}f) - (f,e_0)\, P(\phi e_{-1}) \\ &= W_{\phi e_{-1}}(f) - (f,e_0)\, P(\phi e_{-1}). \end{aligned}$$

Hence $W_\phi W_{e_{-1}} - W_{\phi e_{-1}}$ is a one dimensional operator. We now proceed inductively. Suppose $W_\phi W_{e_{-n}} - W_{\phi e_{-n}}$ has been shown to be compact. Then

$W_\phi W_{e_{-n-1}} - W_{\phi e_{-n-1}} = (W_\phi W_{e_{-n}} - W_{\phi e_{-n}})W_{e_{-1}} + (W_{\phi e_{-n}} W_{e_{-1}} - W_{(\phi e_{-n})e_{-1}})$ so

that the latter is compact. Hence the induction is complete and we can conclude that $W_\phi W_p - W_{\phi p}$ is compact for every trigonometric polynomial p. Since such polynomials are dense in $C(T)$ and the map $\phi \to W_\phi$ is isometric, the proof is complete.■

The next step in our programme is to consider the commutator ideal $\mathcal{I}$ in A, i.e. the smallest two sided ideal in A which contains all the commutators TS - ST, for S,T in A. Obviously $A/\mathcal{I}$ is a commutative Banach algebra and by Theorem 6.3.2, it is even a C* algebra. One might expect however that $\mathcal{I}$ would be extremely difficult to work with. Fortunately, this is not the case and the following result brings us onto familiar ground.

(6.4.3) THEOREM

$\mathcal{I}$ is the ideal K(H) of compact operators.

Proof. The Toeplitz operator W_{e_1} maps $f(t)$ onto $tf(t)$, $f \in H$. If we consider the action of this operator on the sequences of Fourier coefficients $\{f_n\}_0^\infty$ we obtain the unilateral shift S on $\ell_2(Z^+)$

$$S : (f_0,f_1,f_2,\ldots) \to (0,f_0,f_1,f_2,\ldots).$$

An easy calculation shows that $S^*S - SS^*$ is one dimensional so that the same is true of $W^*_{e_1}W_{e_1} - W_{e_1}W^*_{e_1}$. Hence $\mathcal{I}$ contains a nonzero finite dimensional operator. Moreover A is an irreducible algebra; indeed the single operator S has no nontrivial reducing subspaces [31]. Hence by Theorem 6.3.3, we know that $A \supset K(H)$.

Next consider the canonical map $\pi : A \to A/K(H)$. Then since $W^*_{e_1}W_{e_1} - W_{e_1}W^*_{e_1}$ is finite dimensional we see that $\pi(W_{e_1})$ is normal. But W_{e_1} generates A, so that $\pi(A)$ is commutative. But this means that $\pi(C) = 0$ whenever C is a commutator in A and hence $\pi(\mathcal{I}) = \{0\}$. But K(H) is the kernel of π so that we conclude that $K(H) \supset \mathcal{I}$.

Finally, since H is separable, we know from Theorem 5.2.1, that $\mathcal{I} \supset K(H)$ so the proof is complete.■

We now consider the commutative C* algebra $A/\mathcal{I}$. We know that all such algebras are isometrically * isomorphic to an algebra $C(\Omega)$. It is hardly surprizing that we can obtain Ω equal to T.

(6.4.4) THEOREM

$A/\mathcal{I}$ is isometrically * isomorphic to $C(T)$, the isomorphism being given by $W_\phi + \mathcal{I} \to \phi$.

Proof. The map $C(T) \to A/I$ given by $\phi \to W_\phi + I$ is clearly linear and, by Lemma 6.4.2, it is multiplicative. Moreover it is clearly surjective. The fact that $W_{\bar{\phi}} = W^*_\phi$ confirms that we have a $*$ isomorphism. The only part of the proof which is not obvious is the fact that we have an isometry. In fact, we can show that $||W_\phi + K|| = ||W_\phi||$ for each $\phi \in C(T)$ and each $K \in I$. Since certain additional facts about $H_2(T)$ seem to be needed for the proof of this fact we refer the reader to [19, p. 180] for further details.■

We are now ready to discuss the problem of invertibility for W_ϕ. The first step is the following:

(6.4.5) LEMMA

If W_ϕ is a Fredholm operator of index zero, then W_ϕ is invertible.

Proof. If W_ϕ were a noninvertible Fredholm operator of index zero, then $N(W_\phi)$ and $N(W^*_\phi)$ would both contain non-zero vectors. Let $W_\phi f = 0$ and $W^*_\phi g = 0$ for f and g nonzero in H. Then we have $P(\phi f) = 0$ and $P(\bar{\phi} g) = 0$ since $W^*_\phi = W_{\bar{\phi}}$. Hence ϕf has a Fourier series with no terms in e^{int}, $n \geq 0$ so that $\overline{\phi f}$ must be in $H_2(T)$; similarly $\phi \bar{g}$ is in $H_2(T)$. Now $f\phi\bar{g}$ is in $L_1(T)$ since it is the product of L_2 functions and since f and $\phi\bar{g}$ are in $H_2(T)$, a simple argument about Fourier series makes it clear that $f\phi\bar{g}$ has a Fourier series without terms involving e^{int}, $n \leq 0$. But applying the same argument to $g\overline{\phi f}$, we have obtained an L_1 function $\psi = f\phi\bar{g}$ such that neither ψ nor $\bar{\psi}$ has terms in e^{int}, $n \leq 0$, in their Fourier series. Hence $\psi = 0$, i.e. $f\phi\bar{g} = 0$. Now ϕ cannot be identically zero if W_ϕ is Fredholm. Hence $f\bar{g}$ is zero on a set of nonzero measure on T. We now need a theorem due to F. and M. Riesz [19, p. 154]: any nonzero function in $H_2(T)$ can vanish only on a set of measure zero. Clearly this would imply that $f\bar{g}$ can likewise vanish only on a set of measure zero and this provides the required contradiction.■

We can now state the main result about invertibility.

(6.4.6) THEOREM

If ϕ is in $C(T)$, then W_ϕ is invertible if and only if $\phi(t) \neq 0$ for all t in T and the winding number of its graph relative to the origin is zero.

Proof. From the previous lemma, W_ϕ is invertible if and only if it is Fredholm of index zero. But from Theorem 6.3.4, since A is a C^* algebra, W_ϕ is a Fredholm operator on H if and only if its image $\pi(W_\phi)$ in $A/K(H)$ is invertible. By Theorems 6.4.3 and 6.4.4, this is true if and only if ϕ is invertible in $C(T)$, i.e. $\phi(t) \neq 0$ on T.

We now identify index with winding number. To do this, suppose ϕ and ψ are continuous non-vanishing functions on T with graphs which are homotopic in $\mathbb{C} - \{0\}$. Let $\Phi : [0,1] \times T \to \mathbb{C} - \{0\}$ effect this homotopy, i.e. $\Phi(0,t) = \phi(t)$, $\Phi(1,t) = \psi(t)$, and $\Phi(s,t) \in \mathbb{C} - \{0\}$ for all $s \in [0,1]$, $t \in T$. Then for each $s \in [0,1]$, the corresponding Toeplitz operator $W_{\Phi(s,\cdot)}$ is Fredholm and the mapping $s \to \kappa(W_{\Phi(s,\cdot)})$ is a continuous integer valued function on $[0,1]$. Therefore $W_{\Phi(s,\cdot)}$ has constant index; in particular W_ϕ and W_ψ have the same index. Suppose ϕ has winding number n relative to the origin. Choose $\psi(t) = e_n(t) = e^{int}$ so that ϕ and ψ are homotopic. But $W_{e_n} = (W_{e_1})^n$ and W_{e_1}, the unilateral shift, has index -1. Hence by the index theorem $\kappa(W_{e_n}) = -n$ and hence W_ϕ has index $-n$. This completes the proof.■

REMARKS

(1) There is a vector valued version of the above result. Suppose all the functions involved take their values in a finite dimensional Hilbert space h, then the index of a Fredholm Toeplitz operator W_ϕ is equal to minus the winding number of the curve $\det \phi(t)$ about the origin [19, p. 59].

(2) Returning to the scalar valued case, it is clear that the spectrum of W_ϕ consists of the graph of ϕ together with certain bounded components of the complement so that the spectrum is a connected set.

(3) Considerable effort has been made to extend the above results to a larger class of functions ϕ. If one considers the same argument applied to $L_\infty(T)$ instead of $C(T)$ it is possible to show that the commutator ideal properly contains $K(H)$ and that the spectrum of W_ϕ is contained in the closed convex hull of the essential range $\{\lambda \in \mathbb{C}$: for each $\varepsilon > 0$, $|\phi(t) - \lambda| < \varepsilon$ is satisfied on a set of positive measure$\}$. The essential range is, in fact, the spectrum of ϕ in the Banach algebra $L_\infty(T)$. A deep result of H. Widom proves that $\sigma(W_\phi)$ is again a connected set. Much

is also known about other classes of symbols; the reader is referred to [19] for details.

6.5 Fredholm theory in von Neumann algebras

Among the C* algebras, a special class has been the subject of intense study for the past 35 years. This class consists of those C* algebras A which are weakly closed, i.e. if T_n is a sequence in A and there exists T in $B(H)$ such that $(T_n x,y) \to (Tx,y)$ for all x,y in H, then $T \in A$. Such algebras are known as <u>von Neumann algebras</u> or <u>W* algebras</u>. An alternative definition involves the notion of <u>commutant</u>; if A is any subset of $B(H)$ we write A' for the commutant $\{T \in B(H) : TS = ST \text{ for every } S \in A\}$. The double commutant A'' is defined as $(A')'$ and it is clear that $A \subset A''$. It is a useful and interesting fact that the W* algebras are precisely those for which $A = A''$, [17, p. 42].

We propose to give a brief exposition of those parts of the theory which we shall subsequently need. Additional information can be found in the standard sources [17], [63]. The programme of study of W* algebras has centered around the decomposition of such an algebra into especially simple W* algebras called <u>factors</u>. A factor is a von Neumann algebra with trivial center, i.e. $A' \cap A$ consists of multiples of the identity. Factors can be classified into three distinct types. For any algebra A, write $P(A)$ for the orthogonal projections which it contains. If E and $F \in P(A)$, write $E \leq F$ if $R(E) \subseteq R(F)$. A projection E in $P(A)$ is called <u>minimal</u> if $E \neq 0$ and $E \geq F$ for F in $P(A)$ implies either $F = 0$ or $F = E$. A factor A is called <u>Type I</u> if $P(A)$ contains at least one minimal projection; it is called <u>Type II</u> if $P(A)$ contains no minimal projections and there exist $E, F \in P(A)$ such that no partial isometry in A maps $R(E)$ isometrically onto $R(F)$. All other factors are called <u>Type III</u>.

We shall be especially concerned with Type II factors A and wish to give a sketch of how a dimension function can be introduced on $P(A)$. For convenience we introduce some notation: if $R(E)$ can be mapped isometrically onto $R(F)$ by some partial isometry in A, we write $E \sim F$; if there exists $F_1 \in P(A)$ such that $F_1 \leq F$ and $E \sim F_1$, we write $E \lesssim F$.

If $E \in P(A)$ and $F_1 \leq E$ such that $F_1 \sim E - F_1$, we write $F_1 = \frac{1}{2} E$. Similarly if $F_2 \leq F_1$ and $F_2 \sim F_1 - F_2$ we write $F_2 = \frac{1}{4} E$. By induction we can define $2^{-n}E$. It is certainly not clear that $2^{-n}E$ need exist. However, suppose E is finite, i.e. $E \geq F$ and $E \sim F$ together imply $E = F$. Then in Type II factors, a Zorn's lemma argument combined with the non existence of minimal projections, shows that $2^{-n}E$ exists for finite projections E [63, p. 74]. Moreover if F is any projection in A, it can be written as $F = \sum_1^\infty F_n$ where the F_n are mutually orthogonal projections in A and $F_n \sim 2^{-k(n)}E$ for some integer $k(n)$. Now we define $\dim F = \Sigma 2^{-k(n)}$.

Detailed examination of the above dimension function shows that it has the following properties:

(a) $0 \leq \dim F \leq \infty$; $\dim F = 0$ if and only if $F = 0$

(b) $\dim F_1 = \dim F_2$ if and only if $F_1 \sim F_2$

(c) $\dim F_1 \leq \dim F_2$ if and only if $F_1 \lesssim F_2$

(d) $\dim F$ is finite if and only if F is a finite projection

(e) $\dim(F_1 + F_2) = \dim F_1 + \dim F_2$ if $F_1F_2 = F_2F_1 = 0$.

Moreover the range of the dimension function is an interval $[0,r]$, $0 \leq r \leq \infty$.

The idea of the dimension function, as we have introduced it, has a strong geometric motivation. However, it is usually neither practical nor desirable to obtain the dimension function for a specific factor in this way. Rather, we shall briefly discuss the notion of trace on a von Neumann algebra, an extension to this general setting of a wellknown concept in finite dimensional vector spaces. Let A denote any von Neumann algebra and write A^+ for the set of positive selfadjoint operators in A. Then a mapping $\phi : A^+ \to [0,\infty]$ is called a trace if

(i) $\phi(S + T) = \phi(S) + \phi(T)$ for S and T in A^+

(ii) $\phi(\lambda S) = \lambda\phi(S)$ for S in A^+ and $\lambda \geq 0$

(iii) $\phi(S) = \phi(U^*SU)$ for S in A^+ and any unitary operator U in A.

If $\phi(S) = 0$ implies $S = 0$, then ϕ is called faithful; if, for each increasing sequence S_n in A^+ with supremum S in A^+, $\phi(S_n) \uparrow \phi(S)$, then ϕ is called <u>normal</u>. The fact which will be of importance to us later is that the dimension function can be obtained by restricting a faithful normal trace onto $P(A)$.

The way is now open for the definition of Fredholm elements in A. If S is a subset of $P(A)$, we define $\sup S$ as the projection onto the subspace spanned by $\cup\{R(E) : E \in S\}$ and $\inf S$ as the projection onto the subspace $\cap\{R(E) : E \in S\}$. For any $T \in A$, define projections

$$N_T = \sup\{E \in P(A) : TE = 0\}$$

$$R_T = \inf\{E \in P(A) : ET = T\}.$$

In the case where $A = B(H)$, N_T and R_T are immediately recognizable as the nullspace and range projections. In the general setting, we define $F(A)$, the <u>finite elements</u> of A to be those T with R_T a finite projection. Then the <u>compact elements</u> $K(A)$ consist of the uniform closure of $F(A)$. Finally, we define <u>Fredholm elements</u> in A to be $\{T \in A : N_T$ is finite and $R(I - E_0) \subseteq R(T)$ for some finite projection $E_0\}$. Now

$$\begin{aligned} N_{T^*} &= \sup\{E \in P(A) : T^*E = 0\} \\ &= \sup\{E \in P(A) : ET = 0\} \\ &= \sup\{E \in P(A) : (I - E)T = T\}. \end{aligned}$$

But if $(I - E)T = T$ then $R(T) \subseteq R(I - E)$ so that, if T is Fredholm, $I - E_0 \leq I - E$ and hence $E \leq E_0$. Thus $N_{T^*} \leq E_0$ and therefore we can define the index for Fredholm elements

$$\kappa(T) = \dim N_T - \dim N_{T^*}.$$

It is worth remarking that the definition of Fredholm elements given above is valid for any von Neumann algebra; Breuer [5] has also extended the notion of index to the general setting. However his index then takes values in a group obtained from the equivalence classes of certain representations of A induced by the finite projections. We also note that, in general, the range of a Fredholm element in A need not be closed. Apart from this feature, the theory proceeds along familiar lines: the compact elements form a closed two sided ideal, the Fredholm elements are characterized as the invertibles modulo the compacts, $1 - C$ is Fredholm of index zero when C is a compact element and the index theorem holds. The details of the proofs can be found in Breuer [5] and [6].

In preparation for the next section, we will give an outline of a classic construction, due to von Neumann and Murray ([49], see also [17, pp. 133-135]), of Type II factors. This is usually referred to as the group-measure construction. Let (Z,ν) be a locally compact measure space and let H denote the Hilbert space $L_2(Z,\nu)$. For each f in $L_\infty(Z,\nu)$, we have the multiplication operator M_f in $B(H)$ given by $M_f g = fg$.

Now suppose G is a discrete group which operates on Z from the left, i.e. for each g in G, there is a homeomorphism of Z whose value at z will be denoted simply as gz. Then let $\tilde{H}$ be the Hilbert space $L_2(Z\times G)$ and for $\Phi \in \tilde{H}$, we define two classes of operators:

$$\tilde{M}_f\Phi(z,g) = f(z)\Phi(z,g) \quad \text{for each } f \in L_\infty(Z)$$

and

$$\tilde{T}_h\Phi(z,g) = \Phi(hz,hg) \qquad \text{for each } h \in G.$$

Now a simple calculation shows that $\tilde{M}_{f_1}\tilde{T}_{h_1}\tilde{M}_{f_2}\tilde{T}_{h_2}$ is equal to $\tilde{M}_f\tilde{T}_h$ where $f(z) = f_1(z)\, f_2(h_1 z)$ and $h = h_1h_2$. Therefore the * subalgebra $\mathcal{B}_0$ of $B(\tilde{H})$ generated by $\{\tilde{M}_f : f \in L_\infty(Z)\}$ and $\{\tilde{T}_h : h \in G\}$ consists of the finite sums

$$\sum_1^n \tilde{M}_{f_i}\tilde{T}_{h_i}.$$

Let $\mathcal{B}$ denote the von Neumann algebra generated by $\mathcal{B}_0$. Then under certain conditions $\mathcal{B}$ is a factor. These conditions both have to do with the nature of the action of G on Z

(i) For each $s \neq e$ in G and measurable non-null set $Z' \subset Z$, there exists $Z'' \subset Z'$ such that $Z'' \cap sZ'' = \Phi$.

(ii) G acts ergodically on Z.

Finally, by imposing further conditions, we obtain Type II_∞ factors: let ν be invariant under the action of G, let $\nu(\{z\}) = 0$ for each $z \in Z$ and let $\nu(Z) = \infty$.

6.6 Further applications

The index theory for Fredholm elements of von Neumann algebras has found certain interesting applications, two of which will be described below.

(6.6.1) FREDHOLM THEORY FOR GROUP ALGEBRAS

Earlier in this chapter, reference was made to Fredholm theory in the setting of abstract harmonic analysis. In returning to this topic, we recall the basic setting as follows: let G be a locally compact abelian group with dual group $\hat{G}$. Suppose Σ is a sub-semigroup of $\hat{G}$ such that Σ generates $\hat{G}$. Define $H_2(\Sigma)$ as the class of functions f in $L_2(G)$ such that the Fourier transform $\hat{f}$ has support in Σ. Let A denote the C^* algebra generated by the Toeplitz operators $\{W_\sigma : \sigma \in \Sigma\}$ acting on the Hilbert space $H = H_2(\Sigma)$. Write I for the commutator ideal of A.

Although the relationship between I and $K(H)$ seems to be obscure, it is possible to see that the almost periodic functions $AP(G)$ play a natural role in the Fredholm theory of A relative to I. Among the several equivalent definitions of $AP(G)$ the following is best suited to our needs: $AP(G)$ consists of the sup-norm closure of the "trigonometric polynomials" $\{\Sigma c_i \phi_i : c_i \in \mathbb{C}, \phi_i \in \hat{G}\}$. Since Σ generates $\hat{G}$, every ϕ_i can be written as $\phi_i = \sigma_1^{-1} \tau_i$ for some σ_i, τ_i in Σ. Then the Toeplitz operator corresponding to $\phi = \Sigma c_i \phi_i$ will be

$$W_\phi = \Sigma c_i W_{\sigma_1^{-1}} W_{\tau_i} = \Sigma c_i W^*_{\sigma_i} W_{\tau_i}$$

so that when we take uniform limits of the trigonometric polynomials ϕ, the limits of the corresponding Toeplitz operators W give the algebra A. (The map $\phi \to W_\phi$ is clearly norm decreasing, it is even an isometry.)

As might be expected, the map $AP(G) \to A/I$ given by $\phi \to W_\phi + I$ is an isometric *-isomorphism. Moreover, W_ϕ is Fredholm relative to I if and only if ϕ is bounded away from zero; in such a case, we expect that some geometric aspect of ϕ will provide an index for W_ϕ. In fact, when $G = R$ and $\Sigma = R^+$, each function in $AP(R)$, bounded away from zero has an associated "mean motion"

$$\lim_{T\to\infty} \frac{1}{2T} [\arg f(T) - \arg f(-T)]$$

which is, in a certain sense, an average winding number (if one is prepared to believe that the graph of an AP function is something like the graph of a periodic function, this has obvious geometric significance; in general, however the mean motion is a real number rather than an integer.)

Now it is shown in [10], that the above mean motion does provide an index i_m for the elements of A which are Fredholm relative to $\mathcal{I}$. However these elements are not likely to be Fredholm operators, so one might wonder if we can associate with A some other algebra $\mathcal{N}$ such that i_m equals an index computed by means of some concept of dimension. The main result of [10] is a demonstration of this possibility using a suitable von Neumann algebra and Breuer's notion of index.

Consider the group-measure construction described in the previous section with the measure space being R with Lebesgue measure and the group being R_d, the additive real numbers with the discrete topology. The action of R_d on R consists of translation. It is easy to check that the required conditions are fulfilled so that the von Neumann algebra $\tilde{\mathcal{N}}$ obtained is a Type II_∞ factor. The generators of $\tilde{\mathcal{N}}$ will be written

$$(\tilde{M}_\phi f)(x,t) = \phi(x)f(x,t)$$

$$(\tilde{T}_\lambda f)(x,t) = f(x - \lambda,\ t - \lambda)$$

for $f \in L_2(R{\times}R_d)$, $\phi \in L_\infty(R)$ and $\lambda \in R$.

Suppose now that $\tilde{p}$ denotes the operator $\tilde{M}_\chi$ where χ is the characteristic function of $[0,\infty)$. Then we will define the required von Neumann algebra $\mathcal{N}$ by writing

$$\mathcal{N} = \tilde{p}\ \tilde{\mathcal{N}}\ \tilde{p}$$

so that $\mathcal{N}$ can be considered as an algebra on $L_2(R^+{\times}R_d)$.

Next, let $\tilde{A}$ denote the C^* algebra on $L_2(R)$ generated by the multiplication operators $M_\phi : f \to \phi f$, $\phi \in L_\infty(R)$ and the translation operators $T_\lambda f(x) = f(x - \lambda)$. Let p denote the operator M_χ. Then A is isomorphic to the sub C^*-algebra of $\tilde{A}$ generated by $\{p\ T_\lambda p : \lambda \in R^+\}$; for if W_σ is a generator of A, then $\mathcal{F} W_\sigma \mathcal{F}^{-1}$ is an operator on $L_2(R^+)$ and since $\sigma \in \hat{R}$ there exists $\lambda \in R^+$ such that $\sigma(x) = e^{i\lambda x}$. It is then simple to compute that $\mathcal{F} W_\sigma \mathcal{F}^{-1} = T_\lambda\ p$. Hence we can regard $\tilde{A}$ as imbedded in A.

Now the obvious map $\tilde{\rho} : \tilde{A} \to \tilde{\mathcal{N}}$ given by

$$\tilde{\rho}(\Sigma M_{\phi_j} T_{\lambda_j}) = \Sigma \tilde{M}_{\phi_j} \tilde{T}_{\lambda_j}$$

turns out to be an isometry so we can extend it to an isometric mapping on all of $\tilde{A}$ with range in $\tilde{N}$. The diagram below is an attempt to make this rather elaborate construction clearer (the generators are placed below the corresponding algebra)

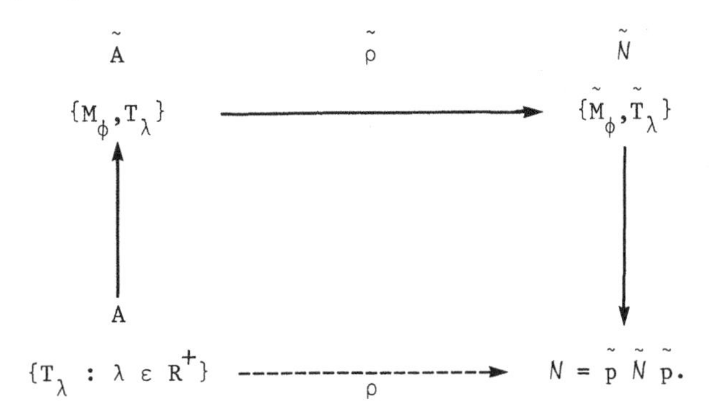

As the diagram indicates, we can define map $\rho : A \to N$ as the composition of the others. Now $\tilde{N}$ is a Type II factor with trace given by

$$\phi(\tilde{M}_\phi \tilde{T}_\lambda) = \begin{cases} 0 & \text{if } \lambda \neq 0 \\ \int_{-\infty}^{\infty} \phi(t)dt & \text{if } \lambda = 0. \end{cases}$$

Also N is a Type II_∞ factor and hence, by Breuer's results, we will have Fredholm elements and an index κ obtained from the dimension function and hence from the trace.

The main result of [10] states that an operator T in A is a Fredholm element relative to I if and only if $\rho(T)$ is Fredholm in the sense of Breuer in the von Neumann algebra N; moreover

$$\kappa(\rho(T)) = i_m(T).$$

We do not propose to go into details of the proofs of these assertions except to exhibit the simple argument which verifies the equality of the two indices.

If W_σ is in A, then we have imbedded A in $\tilde{A}$ and obtained the translation T_λ on $L_2(R^+)$ with $\lambda \geq 0$. Now the AP function corresponding to T_λ is $f(t) = e^{i\lambda t}$ and therefore $i_m(T_\lambda) = -\lambda$. On the other hand, $\rho(T_\lambda)$ is an isometry so $N(\rho(T_\lambda)) = \{0\}$. It remains to consider

$$N(\rho(T_\lambda)^*) = N(\tilde{T}_\lambda^*) = N(\tilde{T}_{-\lambda}).$$

Now $\tilde{T}_{-\lambda}f(x,t) = pf(x + \lambda, t + \lambda) = 0$ if and only if f has its support in the strip $0 < x < \lambda$ in the (x,t) plane. Let $\tilde{\chi}$ be the characteristic function of this strip. Then $N(\tilde{T}_{-\lambda}) = \chi L_2(R^+ \times R_d)$ so that the projection onto $N(\tilde{T}_{-\lambda})$ is $M_{\tilde{\chi}}$ and we compute the trace

$$\phi(M_{\tilde{\chi}}) = \int_{-\infty}^{\infty} \tilde{\chi}(t)dt = \int_0^{\lambda} dt = \lambda.$$

Hence the index $\kappa(\rho(T_\lambda))$ is $\dim N(\rho(T_\lambda)) - \dim N(\rho(T_\lambda)^*) = 0 - \lambda = -\lambda$. The equality of the indices of the generators T_λ can now be extended to equality on the set of Fredholm elements.

Although the above discussion has been restricted to the special case $G = R$, $\sum = R^+$, some extensions to more general groups have been made [10] but so far it has not been possible to attain great generality. The extensions obtained are to groups $H \times V$ where $\hat{H}$ is torsion free and $\hat{V}$ is either R or a subgroup of R_d. It remains to be seen how far these matters can be developed.

(6.6.2) FREDHOLM THEORY AND FINITE DIFFERENCE EQUATIONS

We now turn our attention to recent work by D. G. Schaeffer [59] on a problem involving finite difference equations. The problem has its origins in finite difference approximations to elliptic boundary value problems on the half plane but we refer the reader to the original source for a discussion of this aspect of the work. We will be content in stating the problem as follows: let $[\cdot\ ,\ \cdot]$ denote the usual inner product in R^n and let N be a fixed vector in R^n. Let H denote the half space $\{x \in R^n : [x,N] \geq 0\}$. For any multi-integer $j = (j_1, j_2, \ldots, j_n)$, we define T_j as the translation on functions defined on R^n given by

$$T_j f(x) = f(x + j).$$

Suppose that $Q(D)$ denotes the finite sum $\Sigma c_j T_j$ where the coefficients c_j are fixed complex numbers. Choose $a > 0$ so that $c_j = 0$ if $[j,N] < -a$ and let S denote the "boundary layer",

$$\{x \in R^n : 0 \leq [x,N] \leq a\}.$$

We are also given "boundary conditions" in terms of a difference operator

$$q(x,D) = \Sigma \gamma_j T_j$$

where, now, the coefficients γ_j are bounded measurable functions of $[x,N]$.

The problem to be studied can now be expressed as:

$$(6.6.3) \qquad \begin{cases} Q(D)v(x) = 0 & \text{for} \quad x \in H\backslash S \\ q(x,D)v(x) = g(x) & \text{for} \quad x \in S. \end{cases}$$

We say that $Q(D)$ is <u>properly elliptic</u> if the function $Q(\xi) = \Sigma c_j e^{i[j,\xi]}$ is non-zero for all $\xi \in R^n$ and $\arg Q$ is periodic. The main result is the following theorem.

(6.6.4) THEOREM

Let $Q(D)$ be properly elliptic. Then the following statements are equivalent:

(i) the homogeneous problem (6.6.3) with $g = 0$ has a unique solution $v = 0$.

(ii) For a set of functions g which are dense in $L_2(S)$, problem (6.6.3) has at least one solution in $L_2(H)$.

Proof. We again use the group-measure construction. In this case, we take R with Lebesgue measure for the measure space and the additive group Z^n with the group action being translation $x \to x - [j,N]$ where $j \in Z^n$. Then the corresponding generators of the von Neumann algebra $\tilde{A}$ on $L_2(R\times Z^n)$ are

$$\tilde{M}_\phi f(t,j) = \phi(t)f(t,j)$$

$$\tilde{T}_k f(t,j) = f(t - [k,N], j - k)$$

with $t \in R$, $j \in Z^n$, $\phi \in L_\infty(R)$, $k \in Z^n$.

Again we have trace

$$\phi\Sigma_i \tilde{M}_{\phi_i}\tilde{T}_{k_i} = \int_{-\infty}^{\infty} \phi_0(t)dt \quad \text{where} \quad \tilde{M}_{\phi_0} \quad \text{is the coefficient}$$

of the identity translation $\tilde{T}_0$. In order to get a Type II_∞ factor in this case, it turns out that we need the curve $\{tN : t \in R\}$ to be dense in the n-dimensional torus T^n. Since we require results on $L_2(H)$, it is convenient to replace $\tilde{A}$ by an isomorphic von Neumann algebra A on $L_2(R^n)$ with generators

$$M_\phi f(x) = \phi([x,N])f(x)$$

$$T_k f(x) = f(x - k) \quad \text{with} \quad x \in R^n,\ \phi \in L_\infty(R),\ k \in Z.$$

Let χ_+ denote the characteristic function of $[0,\infty)$ and write E for the projection M_{χ_+}. Then we will be concerned with the von Neumann algebra $\mathcal{B}$ defined as $E\,\mathcal{A}\,E$ which can be considered to act on $L_2(H)$. Now let $\mathcal{S}$ denote the subspace of $L_2(H)$ of functions with support in the boundary layer S. Then $\mathcal{S}$ is the range of the projection M_{χ_a} where χ_a is the characteristic function of $[0,a)$. Hence

$$\dim \mathcal{S} = \operatorname{tr} M_{\chi_a} = \int_{-\infty}^{\infty} \chi_a(t)dt = \int_0^a dt = a.$$

Now suppose we define χ to be the characteristic function of $H\backslash S$. Now $\chi(x)f(x) = (M_{\chi_+} - M_{\chi_a})f(x)$ so that $\chi Q(D) \in \mathcal{B}$. Our main goal is to show that $\chi Q(D)$ is Fredholm with index zero in the sense of Breuer relative to the von Neumann algebra $\mathcal{B}$. Moreover, the range of $\chi Q(D)$ is $L_2(H\backslash S)$.

Suppose that the above two facts were obtained. Then we will indicate how the remaining part of the proof can be deduced. We need one additional fact about von Neumann algebras: suppose X is the range of a projection in $\mathcal{B}$ and A is any operator in $\mathcal{B}^+$. Then from the result

$$\dim X + \dim N(A) = \dim \overline{X + N(A)} + \dim[X \cap N(A)]$$

which is proved in [17, p. 238], we can deduce that

$$\dim X = \dim[\overline{X + N(A)} \ominus N(A)] + \dim[X \cap N(A)]$$

and hence, since $\overline{X + N(A)} \ominus N(A) \sim \overline{AX}$, we obtain

$$\dim X = \dim \overline{AX} + \dim[X \cap N(A)]. \tag{6.6.5}$$

We apply this result with $X = N(\chi Q(D))$ and $A : L_2(H) \to L_2(S)$ being the operator defined by $q(x,D)$.

Suppose then that the homogeneous problem (6.6.3) has only the zero solution. That is equivalent to $X \cap N(A)$ consisting of just $\{0\}$. Then $\overline{AX}$ is a subspace of $L_2(S)$ and from (6.6.5) and the fact that $\chi Q(D)$ is Fredholm with index zero, we get

$$\dim \overline{AX} = \dim \chi = \dim N(\chi Q(D)) = \dim[R(\chi Q(D))]$$
$$= \dim[L_2(H\backslash S)^{\perp}] = \dim L_2(S) = a < \infty.$$

Since the dimension function is obtained from a faithful trace, we deduce that $\overline{AX} = L_2(S)$, i.e. assertion (ii) in the statement of our theorem is valid.

Conversely, if (ii) holds, we have $\overline{AX} = L_2(S)$ so that from (6.6.5) $\dim[X \cap N(A)] = \{0\}$. Again the faithful trace argument leads to the conclusion that $X \cap N(A) = \{0\}$ as required.

It now remains to investigate the proofs of the two assertions made about $\chi Q(D)$. The easier of the two is the fact that $\chi Q(D)$ has range $L_2(H\backslash S)$. We observe immediately that $R(\chi Q(D)) \subset R(X) = L_2(H\backslash S)$. Now suppose $f \in L_2(H\backslash S)$. Then since $Q(D)$ is properly elliptic, $Q(\xi)$ is nonvanishing on the n-dimensional torus T^n so that, by taking Fourier transforms, we see that $Q(D)$ is invertible on $L_2(R^n)$. Therefore there exists $v \in L_2(R^n)$ such that $Q(D)v = f$. If E denotes the projection of $L_2(R^n)$ onto $L_2(H)$, then $\chi Q(D)Ev = \chi Q(D)v = \chi f = f$ because of the way in which S was defined. Thus $R(\chi Q(D)) = L_2(H\backslash S)$.

Finally, we need the fact that $\chi Q(D)$ is a Fredholm element of $\mathcal{B}$ with zero index. Let $C(T^n)$ denote the space of continuous complex valued functions on the n-dimensional torus T^n with the sup-norm. For $\Phi \in C(T^n)$, let $\Phi(D)$ be defined by

$$\Phi(D) = E\mathcal{F}^{-1}\Phi\mathcal{F}E$$

where $\mathcal{F}$ denotes the Fourier transform and E, as before, is the projection of $L_2(R^n)$ onto $L_2(H)$. We claim that $\Phi(D) \in \mathcal{B}$. To see this, suppose $\Phi(\xi) = b_j e^{i[j,\xi]}$; then it is easy to see that $\Phi(D) = E\Sigma b_j T_j E$ so that in this case $\Phi(D) \in \mathcal{B}$. Moreover for any $\Phi \in C(T^n)$, there is a sequence of exponential polynomials converging uniformly to Φ. Since the map $\Phi \to \Phi(D)$ is norm reducing, we can conclude that $\Phi(D) \in \mathcal{B}$ for all Φ in $C(T^n)$. Now the map $\Phi \to \Phi(D)$ is not an algebra homomorphism but we can prove that, if Φ and Ψ belong to $C(T^n)$, then $\Phi\Psi(D) - \Phi(D)\Psi(D) \in \mathcal{K}$, the ideal generated in $\mathcal{B}$ by operators whose range has finite dimension relative to $\mathcal{B}$. For an easy calculation shows that $ET_jET_kE - ET_{j+k}E \in \mathcal{K}$ and hence that the required result holds when Φ and Ψ are exponential polynomials. By an obvious limit argument, we get the general result.

Now consider $Q \varepsilon C(T^n)$, the function associated with the problem (6.6.3) which we are studying. Then since Q is non vanishing, we have $Q^{-1} \varepsilon C(T^n)$ and from the result of the previous paragraph, $Q^{-1}(D)$ is an inverse of $Q(D)$, modulo K. Thus $Q(D)$ is a Fredholm element in B.

To show that $Q(D)$ has index zero, we use the homotopy invariance of the index, observing that $Q(D)$ is homotopic to identity. A function implementing this homotopy is $h(t,\xi) = \exp(t \log Q(\xi))$, $0 \leq t \leq 1$, where $h(0,D) = I$ and $h(1,D) = Q(D)$.■

APPENDIX 1

The notion of the gap between two closed linear subspaces of a Banach space was used in the proofs of Theorems 4.2.1 and 4.2.2. This idea has considerable intrinsic interest. For historical remarks and still further information we refer the reader to the survey article of Gohberg and Krein [27] and to the book of Kato [40].

Let E and F be two closed linear subspaces of the Banach space X. Consider first

$$\alpha(E,F) = \sup\{||x + F|| : x \in E, ||x|| = 1\}$$

where we take this quantity to be zero if $E = \{0\}$. The gap (or opening), $\theta(E,F)$, between E and F is defined to be the maximum of the numbers $\alpha(E,F)$ and $\alpha(F,E)$. It is clear that $0 \leqq \theta(E,F) \leqq 1$, $\theta(E,F) = 0$ if and only if $E = F$ and $\theta(E,F) = \theta(F,E)$.

We shall require the well-known Borsuk-Ulam theorem. This deals with a continuous mapping T defined on the subset S of an $(n + 1)$-dimensional space E, $S = \{x \in E : ||x|| = 1\}$ with values in an n-dimensional space F. The theorem asserts that there exists $x_0 \in S$ where $T(x_0) = T(-x_0)$. For a careful account of this and related results see the book entitled Modern Algebraic Topology by D. G. Bourgin.

THEOREM 1

Suppose that $\theta(E,F) < 1$. Then either E and F are both infinite-dimensional or both are finite-dimensional with the same dimension.

Proof. We may assume that at least one of the subspaces, say F, is finite-dimensional with dimension $n < \infty$. We shall show that if E contains an $(n + 1)$-dimensional subspace E_0, then $\theta(E_0,F) = 1$ which is contrary to our hypothesis. Therefore $\dim E \leqq \dim F$. By symmetry, $\dim F \leqq \dim E$.

Suppose that the subspace E_0 exists. We shall obtain $\theta(E_0,F) = 1$ by finding $x_0 \in E_0$, $||x_0|| = 1 = ||x_0 + F||$. It will be convenient to work in the finite-dimensional space $M = E_0 + F$.

First we treat the case where the norm, $||x||$, on M is strictly convex--that is, if x and y are in M and linearly independent then $||x + y|| < ||x|| + ||y||$. We claim that, for each $x \in M$ there is a unique $y \in F$ such that $||x - y|| = ||x + F||$. To this end we may assume that $x \notin F$. We take a sequence $\{y_k\}$ in F, where $||x - y_k|| \to ||x + F||$. Clearly $\{y_k\}$ is a bounded sequence. Therefore, as F is finite-dimensional there is a subsequence of $\{y_k\}$ converging to some element $y \in F$ and $||x - y|| = ||x + F||$. Suppose also that $z \in F$ and $||x - z|| = ||x + F||$. Then

$$2||x + F|| \leqq ||2x - (y + z)|| \leqq ||x - y|| + ||x - z|| = 2||x + F||.$$

Consequently

$$||(x - y) + (x - z)|| = ||x - y|| + ||x - z||.$$

Therefore $x - y$ and $x - z$ are linearly dependent. Then for a scalar c we get $x - y = c(x - z)$ and this implies that $(1 - c)x = y - cz$. Inasmuch as $x \notin F$ this shows that $c = 1$ and, therefore, $y = z$.

Now we consider the well-defined mapping T of M onto F where $T(x)$ is the unique element of F where $||x - T(x)|| = ||x + F||$. Note that $T(-x) = -T(x)$ because

$$||-x - T(-x)|| = ||-x + F|| = ||x + F|| = ||x - T(x)|| = ||-x + T(x)||.$$

Next we show that T is continuous. Suppose that $x_k \to x$ in M. Then

$$||T(x_k) - x_k|| = ||x_k + F|| \to ||x + F|| = ||T(x) - x||$$

and since

$$||T(x_k)|| \leqq ||T(x_k) - x_k|| + ||x_k||$$

we see that $\{T(x_k)\}$ is a bounded sequence in F. Therefore there is a subsequence $\{x_{k_j}\}$ of $\{x_k\}$ and $z \in F$ where $T(x_{k_j}) \to z$. But then

$$||x - z|| = \lim||x_{k_j} - T(x_{k_j})|| = \lim||x_{k_j} + F|| = ||x + F||$$

and this shows that $z = T(x)$.

Now consider T as a continuous mapping of the set $S = \{x \in E_0 : ||x|| = 1\}$ into F. The Borsuk-Ulam theorem gives some $x_0 \in S$ where $T(x_0) = T(-x_0)$. Recall that $T(-x_0) = -T(x_0)$. Therefore $T(x_0) = 0$. But then

$$||x_0 + F|| = ||x_0 - T(x_0)|| = ||x_0|| = 1.$$

This shows that $\theta(E_0,F) = 1$ as desired.

Now we treat the case where the norm on M is not necessarily strictly convex. Let $|x|$ be any strictly convex norm on M, say the Euclidean norm. For each positive integer k, define a norm $||x||_k$ on M by

$$||x||_k = ||x|| + k^{-1}|x|.$$

These norms have the advantage over the original norm $||x||$ of being strictly convex. Therefore, by the above analysis, there is, for each k, an element $x_k \in E_0$ where

$$||x_k||_k = 1 = ||x_k + F||_k.$$

Clearly $||x_k|| \leqq 1$. Therefore there is a subsequence $\{x_{k_j}\}$ and $y \in E_0$, $||y|| \leqq 1$, so that $x_{k_j} \to y$. Now

$$1 = ||x_{k_j} + F||_{k_j} = ||x_{k_j} + F|| + (k_j)^{-1}|x_{k_j} + F|.$$

The first term on the right approaches $||y + F||$ and the second term approaches zero as $j \to \infty$. Therefore $||y + E|| = 1$. Since $||y|| \leqq 1$ we get $||y|| = 1$ and $\theta(E_0,F) = 1$ as desired.■

We recall some notation from Chapter 1. If E is a subspace of a Banach space X, then

$$E^{\perp} = \{x^* \in X^* : x^*(x) = 0 \text{ for all } x \in E\}.$$

For a subspace W of X (or X^*) we use $S(W)$ to denote the set of elements of W of norm one.

LEMMA

Let E be a subspace of a Banach space X, $x_0 \in X$ and $x_0^* \in X^*$. Then

(a) $||x_0^* - E^{\perp}|| = \sup\{|x_0^*(x)| : x \in S(E)\}$

(b) $||x_0 - E|| = \sup\{|x^*(x_0)| : x^* \in S(E^{\perp})\}$.

Proof. The arguments of Lemma 1.2.5 suffice to show (a). To see (b) consider first $y \in E$ and $x^* \in S(E^{\perp})$. Then

$$|x^*(x_0)| = |x^*(x_0 - y)| \leqq ||x - y||.$$

Therefore $||x_0 - E||$ majorizes the sup of (b). On the other hand, by a corollary to the Hahn-Banach theorem, there exists $x^* \in E^\perp$ where $||x^*|| = 1$ and

$$|x^*(x_0)| = ||x_0 - E||.$$

This shows (b).■

THEOREM 2

Let E and F be two closed linear subspaces of a Banach space X. Then

$$\theta(E,F) = \theta(E^\perp,F^\perp).$$

Proof. We use the preceding lemma repeatedly in our argument.

$$\begin{aligned}
\theta(E,F) &= \max\{\sup_{S(F)} ||x - E||, \sup_{S(E)} ||x - F||\} \\
&= \max\{\sup_{S(F)} \sup_{S(E^\perp)} |x^*(x)|, \sup_{S(E)} \sup_{S(F^\perp)} |x^*(x)|\} \\
&= \max\{\sup_{S(E^\perp)} \sup_{S(F)} |x^*(x)|, \sup_{S(F^\perp)} \sup_{S(E)} |x^*(x)|\} \\
&= \max\{\sup_{S(E^\perp)} ||x^* - F^\perp||, \sup_{S(F^\perp)} ||x^* - E^\perp||\} \\
&= \theta(E^\perp,F^\perp).■
\end{aligned}$$

REFERENCES

NOTE: As a convenience to the reader the page(s) in this text on which a paper is referenced are listed after the reference.

1. F. V. Atkinson, The normal solvability of linear equations in normed spaces, Mat. Sbornik 28 (70), (1951), 3-14 (Russian). 17, 66.

2. S. Banach, Théorie des Operations Lineaires, Monografje Matematyczne, Warsaw, 1932.

3. B. A. Barnes, The Fredholm elements of a ring, Canad. J. Math. 21 (1969), 84-95. 103, 107.

4. C. Bessaga and A. Pelczyński, On bases and unconditional convergence of series in Banach spaces, Studia Math. 17 (1958), 151-164. 88.

5. M. Breuer, Fredholm theories in von Neumann algebras I, Math. Ann. 178 (1968), 243-254. 103, 123.

6. M. Breuer, Fredholm theories in von Neumann algebras II, Math. Ann. 180 (1969), 313-325. 103, 123.

7. J. W. Calkin, Two-sided ideals and congruences in the ring of bounded operators in Hilbert spaces, Ann. of Math. (2) 42 (1941), 839-873. 2, 81.

8. S. R. Caradus, Operators of Riesz type, Pacific J. Math. 18 (1966), 61-71. 98.

9. L. A. Coburn and R. G. Douglas, On C*-algebras of operators on a half space I, Inst. Hautes Études Sci. Publ. Math. No. 40 (1971), 59-67. 103, 116.

10. L. A. Coburn, R. G. Douglas, D. G. Schaeffer and I. M. Singer, On C*-algebras of operators on a half space II: Index theory, Inst. Hautes Études Sci. Publ. Math. No. 40 (1971), 69-79. 103, 126, 127, 128.

11. L. A. Coburn, R. G. Douglas and I. M. Singer, An index theorem for Wiener-Hopf operators on the discrete quarter plane, J. Differential Geom. 6 (1972), 587-595. 103.

12. L. A. Coburn and A. Lebow, Algebraic theory of Fredholm operators, J. Math. Mech. 15 (1966), 577-584. 103, 110.

13. L. W. Cohen and N. Dunford, Transformations on sequence spaces, Duke Math. J. 3 (1937), 689-701. 76.

14. T. Crimmins and P. Rosenthal, On the decomposition of invariant subspaces, Bull. Amer. Math. Soc. 73 (1967), 97-99. 51.

15. M. M. Day, Normed Linear Spaces, Springer-Verlag, Berlin, 1962.

16. A. Devinatz, On Wiener-Hopf operators, Functional Analysis (Proc. Conf. Irvine, California 1966), 81-118. 116.

17. J. Dixmier, Les Algèbres d'Operateurs dans l'espace Hilbertien, Ganthier-Villars, Paris, 1969.

18. J. Dixmier, Les C*-algebras et leurs Representations, Ganthier-Villars, Paris, 1969.

19. R. G. Douglas, Banach Algebra Techniques in Operator Theory, Academic Press, New York, 1972.

20. R. G. Douglas, Banach Algebra Techniques in the Theory of Toeplitz Operators, Regional Conference Series in Mathematics No. 15, Amer. Math. Soc., Providence, Rhode Island. 103.

21. N. Dunford and B. J. Pettis, Linear operators on summable functions, Trans. Amer. Math. Soc. 47 (1940), 323-392. 33.

22. N. Dunford and J. T. Schwartz, Linear Operators Part I, Interscience Publishers, Inc., New York.

23. P. Enflo, A counterexample to the approximation problem in Banach spaces, Acta. Math. 130 (1973), 309-317. 77.

24. T. A. Gillespie and T. T. West, A characterization and two examples of Riesz operators, Glasgow Math. J. 9 (2), 106-110. 58.

25. I. C. Gohberg, On linear equations in normed spaces, Dokl. Akad. Nauk SSSR, 76 (1951), 477-480 (Russian). 17.

26. I. C. Gohberg, On linear equations depending analytically on a parameter, Dokl. Akad. Nauk SSSR 78 (1951), 629-632 (Russian). 17.

27. I. C. Gohberg and M. G. Krein, The basic propositions on defect numbers, root numbers and indices of linear operators, Uspekhi Math. Nauk SSSR 12, (2) 74 (1957), 43-118 (Russian). Amer. Math. Soc. Transl. (2) 13 (1960), 185-265. 17, 61, 133.

28. I. C. Gohberg, A. S. Markus and I. A. Fel'dman, Normally solvable operators and ideals associated with them, Bul. Akad. Stiince Rss Moldoven. 10 (76) (1960), 51-69 (Russian). Amer. Math. Soc. Transl. (2) 61 (1967), 63-84. 81, 101.

29. S. Goldberg, Unbounded Linear Operators with Applications, McGraw-Hill, New York, 1966.

30. B. Gramsch, Ein schema zur theorie Fredholmschen endomorphismen und eine andwendung auf die idealkette der Hilbertraumen, Math. Ann. 171 (1967), 263-272. 103, 108, 110.

31. P. R. Halmos, A Hilbert Space Problem Book, Van Nostrand, 1967.

32. R. H. Herman, On the uniqueness of the ideals of compact and strictly singular operators, Studia Math. 29 (1968), 161-165. 81.

33. H. Heuser, Uber operaten mit endlich defekten, Inaug. Diss., Tubingen, 1956. 57.

34. K. Hoffman, Banach Spaces of Analytic Functions, Prentice-Hall, 1962.

35. M. A. Kaashoek, Ascent, descent nullity and defect, a note on a paper by A. E. Taylor, Math. Ann. 172 (1967), 105-116. 56.

36. M. A. Kaashoek and D. C. Lay, On operators whose Fredholm set is the complex plane, Pac. J. Math. 21 (1967), 275-278. 55.

37. M. I. Kadéc, Linear dimension of the spaces L_p and l_q, Uspehi Mat. Nauk 13 (1958), 95-98 (Russian). 101.

38. S. Kaniel and M. Schechter, Spectral theory for Fredholm operators, Comm. Pure Appl. Math. 16 (1963), 423-448. 55.

39. T. Kato, Perturbation theory for nullity, deficiency and other quantities of linear operators, J. Analyse Math. 6 (1958), 261-322. 56.

40. T. Kato, Perturbation Theory for Linear Operators, Springer-Verlag, Berlin, 1966.

41. D. Kleinecke, Almost-finite, compact, and inessential operators, Proc. Amer. Math. Soc. 14 (1963), 863-868. 33.

42. M. G. Krein, Integral equations on a half-line with kernel depending upon the difference of the arguments, Uspeki Mat. Nauk 13 (1958), no. 5 (83), 3-120 (Russian). Amer. Math. Soc. Transl. (2) 22 (1962). 115.

43. M. G. Krein, M. A. Krasnosel'skii and D. C. Mil'man, On the defect of linear operators in Banach space and on some geometric problems, Sbornik Trud. Inst. Mat. Akad. Nauk Ukr. SSR 11 (1948), 97-112 (Russian). 55.

44. D. C. Lay, Spectral analysis using ascent, descent, nullity and defect, Math. Ann. 184 (1970), 197-214. 56.

45. A. Lebow and M. Schechter, Semigroups of operators and measures of non-compactness, J. Funct. Anal. 7 (1971), 1-26. 55, 70, 73, 95.

46. J. Lindenstrauss and L. Tzafriri, Classical Banach Spaces, Lecture Notes in Mathematics 338, Springer-Verlag, 1973. 77.

47. J. Lindenstrauss, Extension of Compact Operators, Mem. Amer. Math. Soc. 48 (1964). 76.

48. I. Maddaus, On completely continuous linear transformations, Bull. Amer. Math. Soc. 44 (1938), 279-282. 76.

49. F. J. Murray and J. von Neumann, On rings of operators, Ann. Math. 37 (1936), 116-229. 124.

50. B. Noble, Methods Based on the Wiener-Hopf Technique, Pergamon Press, 1958. 115.

51. R. Paley, Some theorems on abstract spaces, Bull. Amer. Math. Soc. 42 (1936), 235-240. 102.

52. W. E. Pfaffenberger, On the ideals of strictly singular and inessential operators, Proc. Amer. Math. Soc. 25 (1970), 603-607. 100.

53. H. Porta, Two-sided ideals of operators, Bull. Amer. Math. Soc. 75 (1969), 599-602. 77.

54. C. E. Rickart, General Theory of Banach Algebras, Princeton, Van Nostrand, 1960.

55. F. Riesz, Über lineare funktionalgleichungen, Acta. Math. 41 (1918), 71-98. 2, 3, 15.

56. F. Riesz and B. Sz-Nagy, Functional Analysis (English translation), Frederick Ungar, New York, 1955.

57. A. P. Robertson and W. Robertson, Topological Vector Spaces, Cambridge University Press, Cambridge, 1964.

58. P. Saphar, Contribution a l'étude des applications linéaires dans un espace de Banach, Bull. Soc. Math. France 92 (1964), 363-384. 56.

59. D. G. Schaeffer, An application of von Neumann algebras to finite difference equations, Ann. Math. 95 (1972), 116-129. 128.

60. J. Schauder, Über lineare, vollstetige funkional operationen, Studia Math 2 (1930), 183-196. 4, 81.

61. M. Schechter, Basic theory of Fredholm operators, Ann. Sci. Norm Sup. Pisa, Sci. fis. mat., III Ser. 21 (1967), 261-280. 55.

62. M. Schechter, Riesz operators and Fredholm perturbations, Bull. Amer. Math. Soc. 74 (1968), 1139-1144. 17, 70, 73.

63. J. T. Schwartz, W* Algebras, Gordon and Breach, New York, 1967.

64. A. E. Taylor, Introduction to Functional Analysis, Wiley, New York, 1961.

65. A. E. Taylor, Theorems on ascent, descent, nullity and defect of linear operators, Math. Ann. 163 (1966), 18-49. 56.

66. T. T. West, The decomposition of Riesz operators, Proc. Lond. Math. Soc. (3), 16 (1966), 737-752. 58.

67. R. J. Whitley, Strictly singular operators and their conjugates, Trans. Am. Math. Soc. 113 (1964), 252-261. 100, 102.

68. B. Yood, Properties of linear transformations preserved under addition of a completely continuous transformation, Duke Math. J. 18 (1951), 599-612. 14, 17, 63, 66.

69. B. Yood, Difference algebras of linear transformations on a Banach space, Pac. J. Math. 4 (1954), 615-636. 33, 95.

70. M. Zippin, On perfectly homogeneous bases in Banach spaces, Israel J. Math. 4 (1966), 265-272. 95.

SYMBOL INDEX

SUBJECT INDEX

M

N

O

P

Q

R

S

T

V

W

For Product Safety Concerns and Information please contact our EU
representative GPSR@taylorandfrancis.com Taylor & Francis Verlag GmbH,
Kaufingerstraße 24, 80331 München, Germany

Printed and bound by CPI Group (UK) Ltd, Croydon, CR0 4YY
01/05/2025
01858595-0001